CAMBRIDGE LIBRARY COLLECTION

Books of enduring scholarly value

Life Sciences

Until the nineteenth century, the various subjects now known as the life sciences were regarded either as arcane studies which had little impact on ordinary daily life, or as a genteel hobby for the leisured classes. The increasing academic rigour and systematisation brought to the study of botany, zoology and other disciplines, and their adoption in university curricula, are reflected in the books reissued in this series.

The Subtropical Garden

The Irish-born gardener and writer William Robinson (1838–1935) travelled widely to study gardens and gardening in Europe and America. In 1871 he founded a weekly illustrated periodical, *The Garden*, which he owned until 1919, and he published numerous books on different aspects of horticulture. Topics included annuals, hardy perennials, alpines and subtropical plants, as well as accounts of his travels. High Victorian garden fashion involved formal beds of exotic and hothouse flowers. Robinson was influential in introducing less formal garden designs, using plants more suited to the English climate. This work was published in 1871, and showed how impressive outdoor displays could be achieved from hardier species, rather than relying on expensive greenhouses for short-lived plants. Robinson's most famous books, *The Wild Garden* (1870) and *The English Flower Garden* (1883) are also reissues in this series.

Cambridge University Press has long been a pioneer in the reissuing of out-of-print titles from its own backlist, producing digital reprints of books that are still sought after by scholars and students but could not be reprinted economically using traditional technology. The Cambridge Library Collection extends this activity to a wider range of books which are still of importance to researchers and professionals, either for the source material they contain, or as landmarks in the history of their academic discipline.

Drawing from the world-renowned collections in the Cambridge University Library, and guided by the advice of experts in each subject area, Cambridge University Press is using state-of-the-art scanning machines in its own Printing House to capture the content of each book selected for inclusion. The files are processed to give a consistently clear, crisp image, and the books finished to the high quality standard for which the Press is recognised around the world. The latest print-on-demand technology ensures that the books will remain available indefinitely, and that orders for single or multiple copies can quickly be supplied.

The Cambridge Library Collection will bring back to life books of enduring scholarly value (including out-of-copyright works originally issued by other publishers) across a wide range of disciplines in the humanities and social sciences and in science and technology.

The Subtropical Garden

Or, Beauty of Form in the Flower Garden

William Robinson

CAMBRIDGE UNIVERSITY PRESS

Cambridge, New York, Melbourne, Madrid, Cape Town,
Singapore, São Paolo, Delhi, Tokyo, Mexico City

Published in the United States of America by Cambridge University Press, New York

www.cambridge.org
Information on this title: www.cambridge.org/9781108037112

This edition first published 1871
This digitally printed version 2011

ISBN 978-1-108-03711-2 Paperback

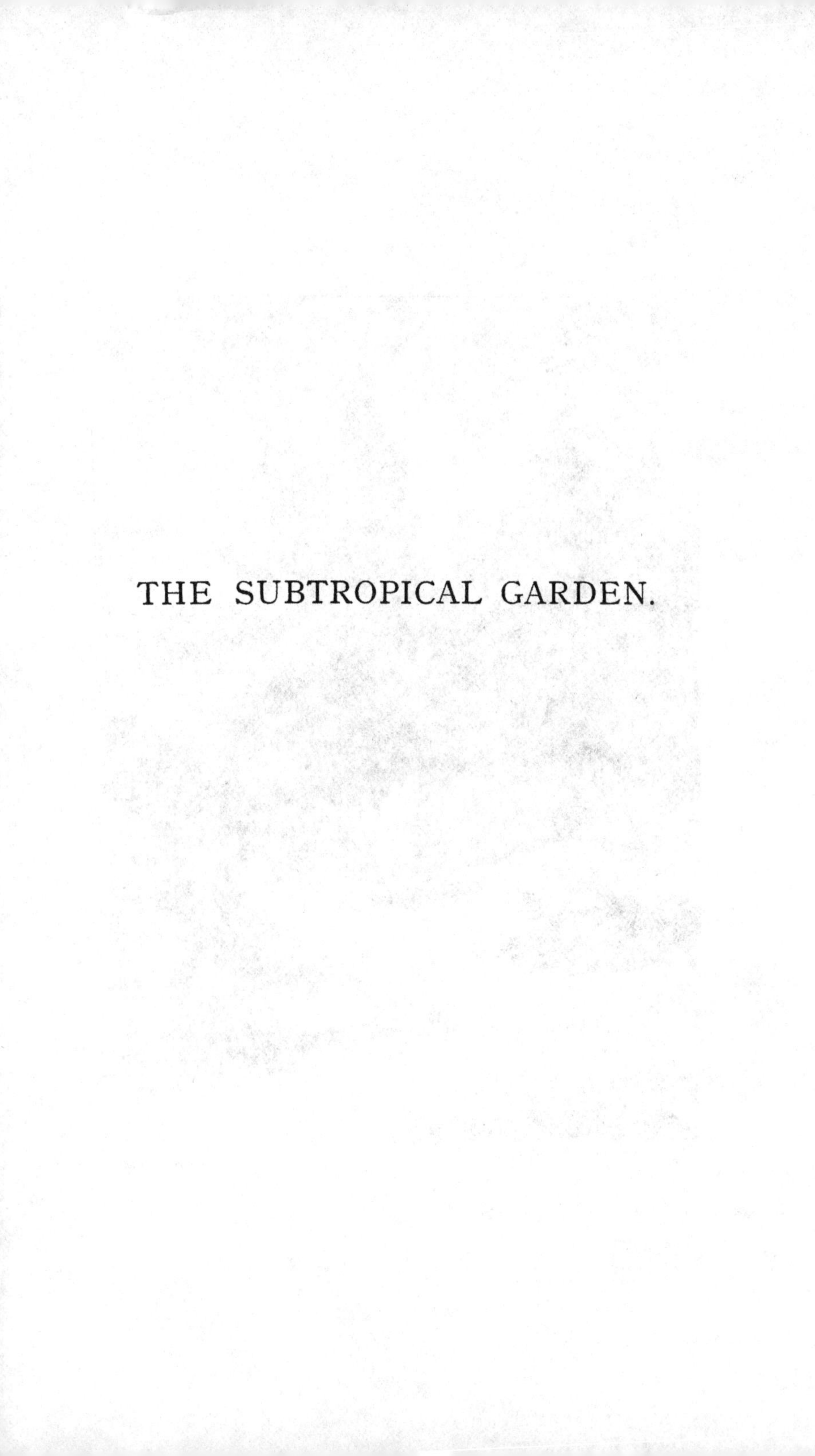

THE SUBTROPICAL GARDEN.

FRONTISPIECE.

THE

SUBTROPICAL GARDEN;

OR,

BEAUTY OF FORM IN THE FLOWER GARDEN.

BY W. ROBINSON, F.L.S,

AUTHOR OF 'ALPINE FLOWERS,' 'THE WILD GARDEN,' 'HARDY FLOWERS,' ETC.

WITH ILLUSTRATIONS.

LONDON:
JOHN MURRAY, ALBEMARLE STREET.
1871.

LONDON:
PRINTED BY WILLIAM CLOWES AND SONS, STAMFORD STREET
AND CHARING CROSS.

PREFACE.

THIS book is written with a view to assist the newly-awakened taste for something more than mere colour in the flower-garden, by enumerating, describing, indicating the best positions for, and giving the culture of, all our materials for what is called "subtropical gardening." This not very happy, not very descriptive name, is adopted from its popularity only; fortunately for our gardens numbers of subjects not from subtropical climes may be employed with great advantage. Subtropical gardening means the culture of plants with large and graceful or remarkable foliage or habit, and the association of them with the usually low-growing and brilliant flowering-plants now so common in our gardens, and which frequently eradicate every trace of beauty of form therein, making the flower-garden a thing of large masses of colour only.

The guiding aim in this book has been the selection of really suitable subjects, and the rejection of many that have been recommended and tried for this purpose. This point is more important than at first sight would appear, for in most of the literature hitherto devoted to the subject plants entirely unsuitable are named. Thus we find such things as Alnus glandulosa aurea and Ulmus campestris aurea (a form of the common elm) enumerated among subtropical plants by one author. Manifestly if these are admissible almost every species of plant is equally so. These belong to a class of variegated hardy subjects that have been in our gardens for ages, and have nothing whatever to do with subtropical gardening. Two other classes have also purposely been omitted: very tender stove-plants, many of which have been tried in vain in the Paris and London Parks, and such things as Echeveria secunda, which though belonging to a type frequently enumerated among subtropical plants, are, more properly, subjects of the bedding class. But if I have excluded many that I know to be unsuitable, every type of the vegetation of northern and temperate countries has been searched for valuable kinds; and as no tropical or subtropical subject that is really effective has been omitted, the result is the most complete selection that is possible from the plants now in cultivation.

No pains have been spared to show by the aid of illustrations the beauty of form displayed by the various types of plants herein enumerated. For some of the illustrations I have to thank MM. Vilmorin and Andrieux, the well-known Parisian firm; for others, the proprietors of the 'Field;' while the rest are from the graceful pencil of Mr. Alfred Dawson, and engraved by Mr. Whymper and

Mr. W. Hooper. I felt that engravings would be of more than their usual value in this book, inasmuch as they place the best attainable result before the reader's eye, thus enabling him to arrange his materials more efficiently. A small portion of the matter of this book originally appeared in my book on the gardens of Paris, in which it will not again be printed. For the extensive list of the varieties of Canna I am indebted to M. Chatè's "*Le Canna.*" Most of the subjects have been described from personal knowledge of them, both in London and Paris gardens.

W. R.

April 3, 1871.

CONTENTS.

PART I.

PAGE

INTRODUCTION AND GENERAL CONSIDERATIONS . . . I

PART II.

DESCRIPTION, ARRANGEMENT, CULTURE, ETC., OF SUITABLE SPECIES, HARDY AND TENDER, ALPHABETICALLY ARRANGED 43

PART III.

SELECTIONS OF PLANTS FOR VARIOUS PURPOSES . 221

LIST OF ILLUSTRATIONS.

Separate plates to face the pages given.

	PAGE
Frontispiece—Hardy and tender Plants in the Sub-tropical Garden.	
Cannas in a London park	13
Anemone japonica alba ...	17
Group and single specimens of plants isolated on the grass	23
Portion of plan showing Yuccas, etc....	25
Formal arrangements in London parks	26
Tree Ferns and other Stove Plants	28
Ailantus and Cannas ...	30
Young Conifers, etc. ...	32
Gourds	34
Section of raised bed at Battersea	40
Acanthus latifolius	47
Aralia canescens	58
Aralia japonica	60
Aralia papyrifera	61
Asplenium Nidus-avis ...	70
Bambusa aurea	72
Bambusa falcata	74
Berberis nepalensis	79
Blechnum brasiliense ...	80
Bocconia cordata	81
Buphthalmum speciosum	83
Caladium esculentum ...	84
Colocasia odorata	85
Canna	86
Carlina acaulis	110
Caryota sobolifera	111
Centaurea babylonica ...	112
Chamædorea	114
Chamærops excelsa ...	116
Cycas	120
Tree Fern	123
Dimorphanthus mandschuricus	124
Erianthus Ravennæ ...	132
Ferula communis	136
Ficus elastica	139
Gynerium argenteum ...	142
Gunnera scabra	144
Heracleum	147
Malva crispa	153
Melianthus major	155
Monstera deliciosa	156
Montagnæa heracleifolia	157
Morina longifolia	158
Mulgedium alpinum ...	159
Musa Ensete	160
Nicotiana Tabacum ...	163
Onopordum Acanthium ...	164
Poa fertilis	174
Rheum Emodi	178
Rhus glabra laciniata ...	180
Seaforthia elegans	185
Solanum robustum	190
Solanum Warscewiczii ...	195
Uhdea bipinnatifida ...	205
Wigandia macrophylla ...	208
Yucca filamentosa	212
Yucca pendula	214
Yucca filamentosa variegata	217

PART I.

INTRODUCTION AND GENERAL CONSIDERATIONS.

SUBTROPICAL GARDENING.

INTRODUCTION AND GENERAL CONSIDERATIONS.

THE system of garden-decoration popularly known as "Subtropical," and which simply means the use in gardens of plants having large and handsome leaves, noble habit, or graceful port, has taught us the value of grace and verdure amid masses of low, brilliant, and unrelieved flowers, and has reminded us how far we have diverged from Nature's ways of displaying the beauty of vegetation, our love for rude colour having led us to ignore the exquisite and inexhaustible way in which plants are naturally arranged. In a wild state brilliant blossoms are usually relieved by a setting of abundant green; and even where mountain and meadow plants of one kind produce a wide blaze of colour at one season, there is intermingled a spray of pointed grass and other leaves, which tone down the mass and quite separate it from anything shown by what is called the "bedding system" in gardens. When we come to examine the most charming examples of our own indigenous or any other wild vegetation, we find that their attraction mainly depends on flower and fern, trailer, shrub, and tree, sheltering, supporting, relieving and beautifying each other, so that the whole array has an indefinite tone, and the mind is satisfied with the refreshing mystery of the arrangement.

We may be pleased by the wide spread of purple on a heath or mountain, but when we go near and examine it in detail, we find that its most exquisite aspect is seen in places where the long moss cushions itself beside the ling, and the fronds of the Polypody peer forth around little masses of heather. Everywhere we see Nature judicious in the arrangement of her highest effects, setting them in clouds of verdant leafage, so that monotony is rarely produced—a state of things which it is highly desirable to attain as far as possible in the garden.

We cannot attempt to reproduce this literally—nor would it be wise or convenient to do so—but assuredly herein will be found the chief source of true beauty and interest in our gardens as well as in those of Nature; and the more we keep this fact before our eyes, the nearer will be our approach to truth and success.

Nature *in puris naturalibus* we cannot have in our gardens, but Nature's laws should not be violated; and few human beings have contravened them more than our flower-gardeners during the past twenty years. We should compose from Nature, as landscape artists do. We may have in our gardens—and without making wildernesses of them either—all the shade, the relief, the grace, the beauty, and nearly all the irregularity of Nature.

Subtropical gardening has shown us that one of the greatest mistakes ever made in the flower-garden was the adoption of a few varieties of plants for culture on a vast scale, to the exclusion of interest and variety, and, too often, of beauty or taste. We have seen how well the pointed, tapering leaves of the Cannas carry the eye upwards; how refreshing it is to cool the eyes in the deep green of those thoroughly tropical Castor-oil plants, with their gigantic leaves; how grand the Wigandia, with its wrought-iron texture and massive outline, looks, after we have surveyed brilliant hues and richly-painted leaves; how greatly the sweeping palm-leaves beautify the

British flower-garden ; and, in a word, the system has shown us the difference between the gardening that interests and delights all beholders, as well as the mere horticulturist, and that which is too often offensive to the eye of taste, and pernicious to every true interest of what Bacon calls the "purest of humane pleasures."

But are we to adopt this system in its purity? as shown, for example, by Mr. Gibson when superintendent of Battersea Park. Certainly not. It is evident, that to accommodate it to private gardens an expense and a revolution of appliances would be necessary, which are in nearly all cases quite impossible, and if possible, hardly desirable. We can, however, introduce into our gardens most of its better features ; we can vary their contents, and render them more interesting by a better and nobler system. The use of all plants without any particular and striking habit, or foliage, or other desirable peculiarity, merely because they are natives of very hot countries, should be tabooed at once, as tending to make much work, and to return—a lot of weeds ; for "weediness" is all that I can ascribe to many Solanums and stove plants, of no real merit, which have been employed under this name. Selection of the most beautiful and useful from the great mass of plants known to science is one of the most important of the horticulturist's duties, and in no branch must he exercise it more thoroughly than in this. Some of the plants used are indispensable—the different kinds of Ricinus, Cannas in great variety, Polymnia, Colocasia, Uhdea, Wigandia, Ferdinanda, Palms, Yuccas, Dracænas, and fine-leaved plants of coriaceous texture generally. A few specimens of these may be accommodated in many gardens; they will embellish the houses in winter, and, transferred to the open garden in summer, will lend interest to it when we are tired of the houses. Some Palms, like Seaforthia, may be used with the best effect for the winter decoration of the conservatory, and be placed out with a good result, and without danger, in summer. Many fine kinds of Dracænas, Yuccas, Agaves, etc., which have been seen to some perfection at our shows of late, are eminently adapted for standing out in summer, and are in fact benefited by it. Among the noblest ornaments of a good conservatory are the Norfolk Island and other tender Araucarias; and these may be placed out for the summer, much to their advantage, because the rains will thoroughly clean and freshen them for winter storing. So with some Cycads and other plants of distinct habit—the very things best fitted to add to the attractions of the flower-garden. Thus we may, in all but the smallest gardens, enjoy all the benefits of what is called Subtropical Gardening, without creating any special arrangements for it.

But what of those who have no conservatory, no hothouses, no means for preserving large tender plants in winter? They too may enjoy the beauty which plants of fine form afford. A better effect than any yet seen in an English garden from tender plants may be obtained by planting hardy ones only! There is the Pampas grass, which when well grown is unsurpassed by anything that requires protection. There are the Yuccas, noble and graceful in outline, and thoroughly hardy, and which, if planted well, are not to be surpassed, if equalled, by anything of like habit we can preserve indoors. There are the Arundos, conspicua and Donax, things that well repay for liberal planting; and there are fine hardy herbaceous plants like Crambe cordifolia, Rheum Emodi, Ferulas, and various graceful umbelliferous plants that will furnish effects equal to any we can produce by using the tenderest exotics. The

Acanthuses too, when well grown, are very suitable for this use. Then we have a hardy Palm, that has preserved its health and greenness in sheltered positions, where its leaves could not be torn to shreds by storms, through all our recent hard winters.

And when we have obtained these, and many like subjects, we may associate them with not a few things of much beauty among trees and shrubs—with elegant tapering young pines, many of which, like Cupressus nutkaensis and the true Thuja gigantea, have branchlets as graceful as a Selaginella; not of necessity bringing the larger things into close or awkward association with the humbler and dwarfer subjects, but sufficiently so to carry the eye from the minute and pretty to the higher and more dignified forms of vegetation. By a judicious selection from the vast number of hardy plants now obtainable in this country, and by associating with them, where it is convenient, house plants that may be placed out for the summer, we may arrange and enjoy charms in the flower-garden to which we are as yet strangers, simply because we have not sufficiently selected from and utilized the vast amount of vegetable beauty at our disposal.

In dealing with the tenderer subjects, we must choose such as will make a healthy growth in sheltered places in the warmer parts of England and Ireland at all events. There is some reason to believe that not a few of the best will be found to flourish much further north than is generally supposed. In all parts the kinds with permanent foliage, such as the New Zealand flax and the hardier Dracænas, will be found as effective as around London and Paris; and to such the northern gardener should turn his attention as much as possible. Even if it were possible to cultivate the softer-growing kinds, like the Ferdinandas, to the same perfection in all parts as in the south of England, it would by no means be everywhere desirable, and especially where expense is a consideration, as these kinds are not capable of being used indoors in winter. The many fine permanent-leaved subjects that stand out in summer without the least injury, and may be transferred to the conservatory in autumn, there to produce as fine an effect all through the cold months as they do in the flower-garden in summer, are the best for those with limited means.

But of infinitely greater importance are the hardy plants; for however few can indulge in the luxury of rich displays of tender plants, or however rare the spots in which they may be ventured out with confidence, all may enjoy those that are hardy, and that too with infinitely less trouble than is required by the tender ones. Those noble masses of fine foliage displayed to us by tender plants have done much towards correcting a false taste. What I wish to impress upon the reader is, that in whatever part of these islands he may live, he need not despair of producing sufficient similar effect to vary his flower-garden or pleasure-ground beautifully by the use of hardy plants alone ; and that the noble lines of a well-grown Yucca recurva, or the finely chiselled yet fern-like spray of a graceful young conifer, will aid him as much in this direction as anything that requires either tropical or subtropical temperature.

Since writing the preceding remarks I have visited America, and when on my way home landed at Queenstown with a view of seeing a few places in the south of Ireland, and among others Fota Island, the residence of Mr. Smith Barry, where I found a capital illustration of what may be easily effected with hardy plants alone. Here an island is planted with a hardy bamboo (*Bambusa falcata*), which thrives so freely as to form great tufts from 16 ft. to 20 ft. high. The result is that the scene reminds one of a bit of the vegetation of the uplands of Java, or that of the bamboo country in China. The thermometer fell last December (1870) seventeen degrees below freezing point, so that they suffered somewhat, but their general effect was not much marred. Accompanying these, and also on the margins of the water, were huge masses of Pampas grass yet in their beauty of bloom, and many great tufts of the tropical-looking New Zealand flax, with here and there a group of Yuccas. The vegetation of the islands and of the margins of the water was composed almost solely of these, and the effect quite unlike anything usually seen in the open air in this country. Nothing in such arrangements as those at Battersea Park equals it, because all the subjects were quite hardy, and as much at home as if in their native wilds. Remember, in addition, that no trouble was required after they were planted, and that the beauty of the scene was very striking a few days before Christmas, long after the ornaments of the ordinary flower-garden had perished. The whole neighbourhood of the island was quite tropical in aspect; and, as behind the silvery plumes of the Pampas grass and the slender wands of the bamboo the exquisitely graceful heads of the Monterey and other cypresses and various pines towered high in the air, it was one of the most charming scenes I have yet enjoyed in the pleasure-grounds of the British Isles. And this, which was simply the result of judiciously planting three or four kinds of hardy plants, will serve to suggest how many other beautiful aspects of vegetation we may create by utilising the rich stores within our reach.

We will next speak of arrangement and sundry other matters of some importance in connection

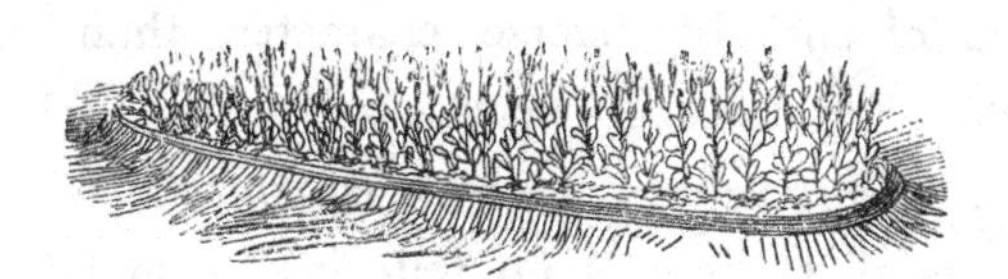

Clumsy mass of Cannas in a London park.

with this subject. The radical fault of the "Sub-tropical Garden," as hitherto seen, is its lumpish monotony and the almost total neglect of graceful combinations. It is fully shown in the London parks every year, so that many people will have seen it for themselves. The subjects are not used to contrast with or relieve others of less attractive port and brilliant colour, but are generally set down in large masses. Here you meet a troop of

Cannas, numbering 500, in one long formal bed—next you arrive at a circle of Aralias, or an oval of Ficus, in which a couple of hundred plants are so densely packed that their tops form a dead level. Isolated from everything else as a rule these masses fail to throw any natural grace into the garden, but, on the other hand, go a long way towards spoiling the character of the subjects of which they are composed. For it is manifest that you get a far superior effect from a group of such a plant as the Gunnera, the Polymnia, or the Castor-oil plant, properly associated with other subjects of entirely diverse character, than you can when the lines or masses of such as these become so large and so estranged from their surroundings that there is no relieving point within reach of the eye. A single specimen or small group of a fine Canna forms one of the most graceful objects the eye can see. Plant a rood of it, and it soon becomes as attractive as so much maize or wheat. No doubt an occasional mass of Cannas, etc., might prove effective—in a distant prospect especially—but the thing is repeated *ad nauseam.*

The fact is, we do not want purely "Subtropical gardens," or "Leaf gardens," or "Colour gardens," but such gardens as, by happy combinations of the materials at our disposal, shall go far to satisfy those in whom true taste has been awakened—and, indeed, all classes. For it is quite a mistake to assume that because people, ignorant of the inexhaustible stores of the vegetable kingdom, admire the showy glares of colour now so often seen in our gardens, they are incapable of enjoying scenes displaying some traces of natural beauty and variety.

The fine-leaved plants have not yet been associated immediately with the flowers; hence the chief fault. Till they are so treated we can hardly see the great use of such in ornamental gardening. Why not take some of the handsomest plants of the medium-sized kinds, place them in the centre of a bed, and then surround them with the gaily-flowering subjects? The Castor-oil plants would not do so well for this, because they are rampant growers in fair seasons, but the Yuccas, Cannas, Wigandias, and small neat Palms and Cycads would suit exactly. Avoid huge, unmeaning masses, and associate more intimately the fine-leaved plants with the brilliant flowers. A quiet mass of green might be desirable in some positions, but even that could be varied most effectively as regards form.

The combinations of this kind that may be made are innumerable, and there is no reason why our beds should not be as graceful as bouquets well and simply made.

However, it is not only by making combinations of the subtropical plants with the gay-flowering ones now seen in our flower-gardens that a beautiful effect may be obtained, but also with those of a somewhat different type. Take, for instance, the stately hollyhock, sometimes grown in such formal plantations as to lose some of its charms, and usually stiff and poor below the flowers. It is easy to imagine how much better a group of these would appear if seen surrounded by a graceful ring of Cannas, or any other tall and vigorous subjects, than they have ever yet appeared in our gardens.

Consider, again, the Lilies, from the superb, tall, and double varieties of the brilliant Tiger lily to the fair White lily or the popular L. auratum. Why, a few isolated heads of Fortune's Tiger lily, rising like candelabra above a group of Cannas, would form one of the most brilliant pictures ever seen in a garden. Then, to descend from a very tall to a very dwarf lily, the large and white trumpet-like flowers of L. longiflorum would look superb, emerging from the outer margin of a mass of subtropical plants, relieved by the rich green within; and anybody, with even a slight knowledge of the lily family, may imagine many other combinations equally beautiful and new. The bulbs would of

Anemone japonica alba. Type of fine-flowered herbaceous plant for associating with foliage-plants.

course require planting in the autumn, and might be left in their places for several years at a time, whereas the subtropical plants might be those that require planting every year; but as the effect is obtained by using comparatively few lilies, the spaces between them would be so large, as to leave plenty of room to plant the others. However, it is worth bearing in mind, that most of the Cannas, by far the finest group of "Subtropical" plants for the British Isles, remain through the winter in beds in the open air protected by litter: hence, permanent combinations of Lilies and Cannas are perfectly practicable.

Then, again, we have those brilliant and graceful hosts of Gladioli, that do not show their full beauty in the florist's stand or in his formal bed, but when they spring here and there, in an isolated manner, from rich foliage, entirely unlike their own pointed sword-like blades. Next may be named the flame-flowered Tritoma, itself almost subtropical in foliage when well grown. Any of the Tritomas furnish a splendid effect grouped near or closely associated with subtropical plants. The lavishly blooming and tropical-looking Dahlia is a host in itself, varying so much as it does from the most gorgeous to the most delicate hues, and differing greatly too in the size of the flowers, from those of the pretty fancy Dahlias to the largest exhibition kinds. Combinations of Dahlias with Cannas and other free-growing subtropical plants have a most satisfactory effect; and where beds or groups are formed of hardy subjects (Acanthuses and the like), in quiet half-shady spots, some of the more beautiful spotted and white varieties of our own stately and graceful Foxglove would be charmingly effective. In similar positions a great Mullein (*Verbascum*) here and there would also suit; while such bold herbaceous genera as Iris, Aster (the tall perennial kinds), the perennial Lupin, Baptisias, Thermopsis, Delphiniums, tall Veronicas, Aconites, tall Campanulas, Papaver bracteatum, Achillea filipendula, Eupatoriums, tall Phloxes, Vernonias, Leptandra, etc., might be used effectively in various positions, associated with groups of hardy subjects. For those put out in early summer, summer and autumn-flowering things should be chosen.

The tall and graceful Sparaxis pulcherrima would look exquisite leaning forth from masses of rich foliage about a yard high; the common and the double perennial Sunflower (*Helianthus multiflorus, fl. pl.*) would serve in rougher parts, where admired; in sheltered dells the large and hardy varieties of Crinum capense would look very tropical and beautiful if planted in rich moist ground; and the Fuchsia would afford very efficient aid in mild districts, where it is little injured in winter, and where, consequently, tall specimens flower throughout the summer months; and lastly, the many varied and magnificent varieties of herbaceous Peony, raised during recent years, would prove admirable as isolated specimens on the grass near groups of fine-foliaged plants. Then again we have the fine Japan Anemones, white and rose, the showy and vigorous Rudbeckias, the sweet and large annual Datura ceratocaula, the profusely-flowering Statice latifolia, the Gaillardias, the Peas (everlasting and otherwise), the ever-welcome African Lily (*Calla*), the handsome Loosestrife (*Lythrum roseum superbum*), and the still handsomer French Willow, and not a few other things which need not be enumerated here, inasmuch as it is hoped enough has been said to show our great and unused resources for adding real grace and interest to our gardens. This phase of the subject—the association of tall or bold flowers with foliage-plants—is so important, that I have bestowed some pains in selecting the many and various subjects useful for it from almost every class of plants; and they will be found in a list at the end of the alphabetical arrangement.

Many charming results may be obtained by carpeting the ground beneath masses of tender subtropical plants with quick-growing ornamental annuals and bedding plants, which will bloom before the larger subjects have put forth their strength and beauty of leaf. If all interested in flower-gardening had an opportunity of seeing the charming effects produced by judiciously intermingling fine-leaved plants with brilliant flowers, there would be an immediate revolution in our flower-gardening, and verdant grace and beauty of form would be introduced, and all the brilliancy of colour that could be desired might be seen at the same time. Here is a bed of Erythrinas not yet in flower: but what affords that brilliant and singular mass of colour beneath them? Simply a mixture of the lighter varieties of Lobelia speciosa with variously coloured and brilliant Portulacas. The beautiful surfacings that may thus be made with annual, biennial, or ordinary bedding plants, from Mignonette to Petunias and Nierembergias, are almost innumerable.

Reflect for a moment how consistent is all this with the best gardening and the purest taste. The bare earth is covered quickly with these free-growing dwarfs; there is an immediate and a charming contrast between the dwarf-flowering and the fine-foliaged plants; and should the last at any time put their heads too high for the more valuable things above them, they can be cut in for a second bloom. In the case of using foliage-plants that are eventually to cover the bed completely, annuals may be sown, and they in many cases will pass out of bloom and may be cleared away just as the large leaves begin to cover the ground. Where this is not the case, but the larger plants are placed thin enough to always allow of the lower ones being seen, two or even more kinds of dwarf plants may be employed, so that the one may succeed the other, and that there may be a mingling of bloom. It may be thought that this kind of mixture would interfere with what is called the unity of effect that we attempt to attain in our flower-gardens. This need not be so by any means; the system could be used effectively in the most formal of gardens.

One of the most useful and natural ways of diversifying a garden, and one that we rarely or never take advantage of, consists in placing really distinct and handsome plants alone upon the grass, to break the monotony of clump margins and of everything else. To follow this plan is *necessary* wherever great variety and the highest beauty are desired in the ornamental garden. Plants may be placed singly or in open groups near the margins of a bold clump of shrubs or in the open grass; and the system is applicable to all kinds of hardy ornamental subjects, from trees downwards, though in our case the want is for the fine-leaved plants and the more distinct hardy subjects. Nothing, for instance, can look better than a well-developed tuft of the broad-leaved Acanthus latifolius, springing from the turf not far from the margin of a pleasure-ground walk; and the same is true of the Yuccas, Tritomas, and other things of like character and hardiness. We may make attractive groups of one family, as the hardiest Yuccas; or splendid groups of one species like the Pampas grass—not by any means repeating the individual, for there are about twenty varieties of this plant known on the Continent, and from these half a dozen really distinct and charming kinds might be selected to form a group. The same applies to the Tritomas, which we usually manage to drill into straight lines; in an isolated group in a verdant glade they are seen for the first time to best advantage: and what might not be

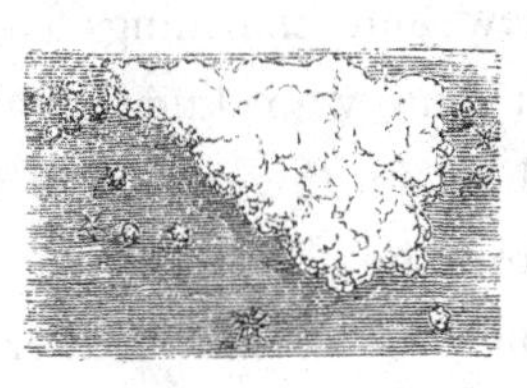

Group and single specimens of plants isolated on the grass.

done with these and the like by making mixed groups, or letting each plant stand distinct upon the grass, perfectly isolated in its beauty!

Let us again try to illustrate the idea simply. Take an important spot in a pleasure-ground—a sweep of grass in face of a shrubbery—and see what can be done with it by means of these isolated plants. If, instead of leaving it in the bald state in which it is often found, we place distinct things isolated here and there upon the grass, the margin of shrubbery will be quite softened, and a new and charming feature added to the garden. If one who knew many plants were arranging them in this way, and had a large stock to select from, he might produce numberless fine effects. In the case of the smaller things, such as the Yucca and variegated Arundo, groups of four or five good plants should be used to form one mass, and everything should be perfectly distinct and isolated, so that a person could freely move about amongst the plants without touching them. In addition to such arrangements, two or three individuals of a species might be placed here and there upon the grass with the best effect. For example, there is at present in our nurseries a great Japanese Polygonum (*P. Sieboldi*), which has never as yet been used with much effect in the garden. If anybody will select some open grassy spot in a pleasure-garden, or grassy glade near a wood—some spot considered unworthy of attention as regards ornamenting it—and plant a group of three plants of this Polygonum, leaving fifteen feet or so between the stools, a distinct aspect of vegetation will be the result. The plant is herbaceous, and will spring up every year to a height of from six feet to eight feet if planted well; it has a graceful arching habit in the upper branches, and is covered with a profusion of small bunches of pale flowers in autumn. It is needless to multiply examples; the plan is capable of infinite variation, and on that account alone should be welcome to all true gardeners.

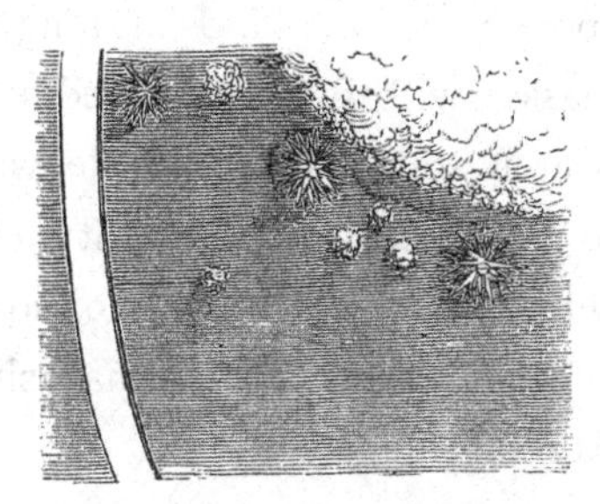

Portion of plan showing Yuccas, Pampas grass, Tritomas, Retinospora, Acanthus latifolius, Arundo Donax variegata, etc., irregularly isolated on the grass.

One kind of arrangement needs to be particularly guarded against — the geometro-picturesque one, seen in some parts of the London parks devoted to subtropical gardening. The plants are very often of the finest kinds and in the most robust health, all the materials for the best results are abundant, and yet the scene fails to satisfy the eye, from the needless formality of many of the beds, produced by the heaping together of a great number of species of one kind in long straight or twisting masses with high raised edges frequently of hard-beaten soil. Many people will not see their way to obliterate the formality of the beds, but assuredly we need not do so to get rid of such effective formality as that shown in the accompanying figure!

Formal arrangements in London parks.

The formality of the true geometrical garden is charming to many to whom this style is offensive; and there is not the slightest reason why the most beautiful combinations of fine-leaved and fine-flowered plants should not be made in any kind of geometrical garden.

But in the purely picturesque garden it is as

needless, as it is in false taste, to follow the course here pointed out. Hardy plants may be isolated on the turf, and may be arranged in beautiful irregular groups, with the turf also for a carpet, or some graceful spray of hardy trailing plants. Beds may be readily placed so that no such objectionable stage-like results will be seen as those shown in the preceding figure: tender plants may be grouped as freely as may be desired—a formal edge avoided by the turf being allowed to play irregularly under and along the margins, while the remaining bare ground beneath the tall plants may be quickly covered with some fast-growing annuals like Mignonette or Nolanas, some soft-spreading bedding plants like Lobelias or Petunias, or subjects still more peculiarly suited for this purpose, such as the common Lycopodium denticulatum and Tradescantia discolor. Choice tender specimens of Tree ferns, etc., placed in dark shady dells, may be plunged to the rims of the pots in the turf or earth, and some graceful or bold trailing herb placed round the cavity so as to conceal it; and in this way such results may be attained as those indicated in the first plate, in those showing the Dimorphanthus, Musa Ensete, and in the frontispiece. The day will come when we shall be as anxious to avoid all formal twirlings in our gardens as we now are to have them perpetrated in them by landscape-gardeners of great repute for applying wall-paper or fire-shovel patterns to the surface of the reluctant earth, and when we shall no more think of tolerating in a garden such a scene as that shown in the preceding figure, than a landscape artist would tolerate it in a picture.

The old landscape-gardening dogma, which tells us we cannot have all the wild beauty of nature in our gardens, and may as well resigr ourselves to the compass, and the level, and the defined daub of colour and pudding-like heaps of shrubs, had some faint force when our materials for gardening were few,* but considering our present rich and, to a great extent, unused stores from every clime, and from almost every important section of the vegetable kingdom, it is demonstrably false and foolish.

To these observations on arrangement, etc., one good rule may be added :—Make your garden as distinct as possible from those of your neighbours—

* "In gardening, the materials of the scene are few, and those few unwieldy, and the artist must often content himself with the reflection that he has given the best disposition in his power to the scanty and intractable materials of nature."—ALLISON.

Shady and sheltered Dell, with Tree Ferns and other Stove Plants placed out for the summer.

which by no means necessitates a departure from the rules of good taste.

I wish particularly to call attention to the fine effects which may be secured, from the simplest and most easily obtained materials, by using some of our hardy trees and shrubs in the subtropical garden. Our object generally is to secure large and handsome types of leaves; and for this purpose we usually place in the open air young plants of exotic trees, taking them in again in autumn; and, perhaps, as we never see them but in a diminutive state, we often forget that, when branched into a large head in their native countries, they are not a whit more remarkable in foliage than many of the trees of our pleasure-grounds. Thus, if the well-known Paulownia imperialis were too tender to stand our winters, and if we were accustomed to see it only in a young and simple-stemmed condition and with large leaves, we should doubtless plant it out every summer as we do the Ferdinanda. There is no occasion whatever to resort to exotic subjects, while we can so easily obtain fine hardy subjects—which, moreover, may be grown by everybody and everywhere. By annually cutting down young plants of various hardy trees and shrubs, and letting them make a clean, simple-stemmed growth every year, we will, as a rule, obtain finer effects than can be got from tender ones. The Ailantus, for example, treated in this way, gives us as fine a type of pinnate leaf as can be desired. Nobody need place Astrapæa Wallichii in the open air, as I have seen done, so long as a simple-stemmed young plant of the Paulownia makes such a column of magnificent leaves. The delicately-cut leaves of the Gleditschias, borne on strong young stems, would be as pretty as those of any fern; and so in the case of various other hardy trees and shrubs. Persons in the coldest and least favourable parts of the country need not doubt of being able to obtain as fine types of foliage as they can desire, by selecting a dozen kinds of hardy trees and treating them in this way. What may be done in this way, in one case, is shown in the accompanying plate, representing a young plant of Ailantus, with its current year's shoot and leaves, standing gracefully in the midst of a bed of Cannas.

A few words may now be added about some types of vegetation which, though not included among what are commonly termed subtropical plants, may yet be judiciously used in combination with them, and go far to produce very charming effects.

AILANTUS AND CANNAS.

Suggesting the effects to be obtained from young and vigorous specimens of hardy fine-leaved trees.

Among conifers we find many subjects of the most exquisite grace, and of a beautiful free and pointed habit, which it is most desirable we should have associated with vegetation more distinguished for brilliancy than grace. They are in many cases as elegantly chiselled and dissected as the finest fern, and it is difficult to find more beautiful masses of verdure than such plants as Retinospora plumosa and R. obtusa display when well developed; they are simply invaluable for those who use them with taste. Apart altogether from our want of a more elegantly diversified surface in the flower-garden—the best and most practical way to meet which is by the use of such plants as these and neat and elegant young specimens of such things as Thujopsis borealis—there is, in many British gardens, a great gulf between the larger tree and shrub vegetation and the humbler colouring material, which most will admit should be filled up, and there is nothing more suitable for it than the many graceful conifers we now possess. Much as conifers are grown with us, how few people have any idea of their great value as ornamental plants for the very choicest position in a garden! We are sometimes too apt to put them in what is called their "proper place,"—or, at all events, too far from the seat of interest to thoroughly enjoy them in winter, when the beauty of their form and their exquisite verdure are best seen. If the dwarfer and choicer conifers were tastefully disposed in and immediately around a flower-garden not altogether spoiled by a profusion of beds for masses of colour, that flower-garden could hardly fail to look as well in winter as in summer; in fact I have seen places where, from rather close association of the more elegant types, the best kind of winter garden was made. Our efforts must tend to prevent a desert-like aspect at any time of the year; and to this end nothing can help us more than a judicious selection of conifers. Almost every beauty of form is theirs. They possess a permanent dignity and interest, always occupying the ground and embellishing it, displaying distinct tints of ever-grateful green in spring and summer, waving majestically before the gusts of autumn, and beautiful when bearing on their deepest green the snows of winter. Some of the more suitable kinds are named in a list at the end of this book, but the graceful pines are so commonly grown that few will have any difficulty in securing proper sorts.

The Gourd tribe is capable, if properly used, of adding much remarkable beauty and character to

Young Conifers and hardy fine-leaved Plants.

the garden; yet, as a rule, it is seldom used. There is no natural order more wonderful in the variety and singular shapes of its fruit than that to which the melon, cucumber, and vegetable marrow belong. From the writhing Snake-cucumber, which hangs down four or five feet long from its stem, to the round enormous giant pumpkin or gourd, the grotesque variation, both in colour and shape and size, is marvellous. There are some pretty little gourds which do not weigh more than half an ounce when ripe; while, on the other hand, there are kinds with fruit almost large enough to make a sponge bath. Eggs, bottles, gooseberries, clubs, caskets, folded umbrellas, balls, vases, urns, small balloons,—all have their likenesses in the gourd family. Those who have seen a good collection of them will be able to understand Nathaniel Hawthorne's enthusiasm about these quaint and graceful vegetable forms when he says: "A hundred gourds in my garden were worthy, in my eyes at least, of being rendered indestructible in marble. If ever Providence (but I know it never will) should assign me a superfluity of gold, part of it shall be expended for a service of plate, or most delicate porcelain, to be wrought into the shape of gourds gathered from vines which I will plant with my own hands. As dishes for containing vegetables they would be peculiarly appropriate. Gazing at them, I felt that by my agency something worth living for had been done. A new substance was born into the world. They were real and tangible existences, which the mind could seize hold of and rejoice in." Of course the climate of New England is much better suited for fully developing the gourd tribe than ours, but it is satisfactory to know that they may be readily and beautifully grown in this country.

There are many positions in gardens in which they might be grown with great advantage; on low trellises, depending from the edges of raised beds, the smaller and medium-sized kinds trained over arches or arched trellis-work, covering banks, or on the ordinary level earth of the garden. Isolated, too, some kinds would look very effective, and in fact there is hardly any limit to the uses to which they might be applied. In the Royal Botanic

Gardens at Dublin, there is a singular wigwam made by placing a number of dead branches so as to form the framework, and then planting Aristolochia Sipho all round these. It runs over them, and the large leaves make a perfect summer roof. A similar tent might be made with the free-growing gourds, and it would have the additional merit of suspending some of the most singular, graceful, and gigantic of all known fruits from the roof. A few words on their culture, and a selection of kinds, occur at the end of the book.

Although some Ferns are named in the descriptive part of this book, it is desirable to allude to the family here. Why do we always put ferns in the shade, when many of the best and hardiest kinds grow freely in the full sun if sufficiently moist at the root? Why do we always confine them to the fernery proper, when there are so many other places that could be graced by their presence? The very highest beauty of form might be added to beds of low flowers, by the introduction of such ferns as the Struthiopteris, Pteris, Lastrea, etc., while they should also be freely planted in various parts of the pleasure-ground, either alone, or grouped with the Acanthuses and other hardy fine-leaved plants. Not a few of the

Umbelliferous plants recommended have foliage as finely cut as any of the Ferns, and would associate very well with them. Even in cases where the soil might not be suitable for ferns, it would, instead of confining them to the fernery proper, be much better to arrange for having small groups or beds of them in places alongside of shady wood-walks or similar positions. By reference to the Osmunda article, it will be seen how these have been grown to magnificent proportions. It may be easily imagined that groups of fine ferns, grown to the luxuriance there described, would contrast with and relieve groups of the brilliant flowers in a superb way.

As the culture of most of the subjects has been sufficiently spoken of in the descriptive part, it is needless to say much of it here, but a few general remarks may help to make the matter clearer to the amateur. It is hoped that the greater number of the hardy subjects enumerated will sufficiently prove that it is not only those persons who have streets of glass-houses to whom the luxury of "subtropical gardening" is accessible. Once placed in suitable soil and position, these hardy kinds may, as a rule, be left to take care of themselves.

A great number of subjects, like the Ricinus

and the Annuals, may be considered practically hardy, inasmuch as they only require to be raised in warm or cool frames, or even (some of them) in the open air. When once planted out for the summer, they give but little further trouble.

In the next group may be placed the tender greenhouse kinds ; long-lived subjects, like the Dracænas, American Aloe, etc., which thrive in greenhouses or conservatories in winter, and are great ornaments there, and which may be placed in the open air in summer without the least injury. Next to the hardy group, this is the most important, from the fact that the subjects are effective at all seasons of the year, and useful indoors as well as without. They also, unlike the following, may be enjoyed by every one who possesses any kind of a cool glazed structure ; and even, in some cases, this is not needed, for I have seen some very fine specimens of Agave americana kept in a large entrance hall in winter, and put out of doors in May to be taken in again in October.

Lastly, we have the least important group of all, and happily also the most costly, viz., those plants which must be kept through the winter and spring in warm stoves, such as Ferdinanda, Solanum, etc. Considering the vast number of hardy and half-hardy plants from which we may select, this type is not worthy of encouragement in gardens generally, with the exception of a few fine things, such as Polymnia grandis. They may, for the sake of convenience, be considered in two sections: those, like the Polymnia, that should be put out in a young state, and which make a fresh and handsome growth during the summer months; and those which, like the Monstera and Anthurium acaule, make no growth whatever during that season. It need not be said that the first section is by far the most important: it comprises the Wigandia, and some of the noblest things used in this way. Plants of the other section can, in the nature of things, be tried in but few places in this country; they are too expensive, and they are not the most effective: but some persons no doubt may take a pleasure in showing what will endure the open air, even if useless for any other purpose. One general rule may be applied to these last-named subjects—they should be allowed to make a strong growth in the hothouse in spring or early summer, and to mature, and, so to speak, harden off that growth before being placed in the open air early in June, or even later if the season be unfavourable.

Speaking generally of all the tender subjects

used, it is necessary to discriminate between kinds that should be planted out in a young state every year, and those which are valuable in proportion to their age and size. Some plants are all the better the higher and larger they are grown; others must be started in a dwarf fresh state every year, or, if not, their foliage will not possess that pristine freshness which charms us when they are properly treated. A large plant of Polymnia grandis, for example, would, if placed in the open air in early summer, speedily become a far from attractive object, while a young plant of the same kind, put out on the same day, would soon produce and carry to the end of the season a mass of fresh and noble leaves. But of course this only applies to kinds that grow rapidly during the summer months in our climate.

With respect to the preparation of the beds for the finer subtropical plants, a peculiar mode is practised in Battersea Park. Here many of the beds are raised above the level of the ground, and underneath and around the mass of light rich soil is a good layer of brick-rubbish, as shown in the accompanying engraving. The soil is first excavated and thrown round the margin of the bed; then the brick-rubbish is put in on the bottom and around the sides also, raising the bed somewhat above the level of the ground; the cavity in the centre is then filled up, generally with fine light rich soil, using as much of the soil that was dug out as is fit to be used, and arranging the remainder round the edge of the raised bed, covering it neatly with turf. The soil may vary in depth from three

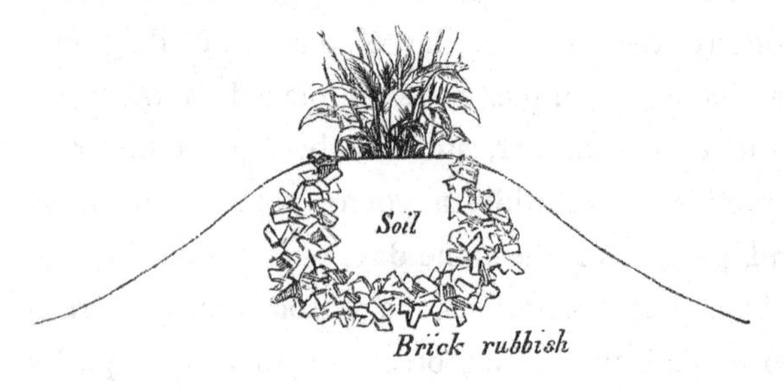

Section of raised bed at Battersea, with brick-rubbish beneath and around the soil.

feet to eighteen inches, according to the kinds of plants to be grown in it. In this way, by presenting a larger surface to the sun, it is considered that a greater amount of heat is obtained; but I certainly think the advantages of the method are not so great in this way as is generally supposed, and that it is quite needless to adopt it in the case of the great majority of subjects. Its chief merit probably is that it secures a better drainage. Good drainage is undoubtedly indispensable, and, still more so, a thoroughly rich and light mass of deep soil, with abundance of water; without these two last conditions it is hopeless to expect a free rich growth, which is the great charm of these plants. Ricinus, Cannas, Ferdinanda, and some of the freer-growing kinds certainly succeed perfectly without any such arrangement as that above described. The more delicate kinds, such as the Solanums and Wigandia macrophylla, would be those most likely to be benefited by it. It is needless to say, that the numerous fine and hardy subjects enumerated in Part II. do not require anything of the kind, although they too will, as a rule, be fine in proportion to the care bestowed in securing for them a deep and rich body of soil.

One most essential matter is the securing of as perfect shelter as is possible. Warm, sunny, and thoroughly sheltered dells should be chosen where convenient; and, in any case, positions which are sheltered should be selected, as the leaves of all the better kinds suffer very much from strong winds, from which they will be protected if judiciously planted near sheltering banks and trees. Even in quite level districts it will be possible to secure shelter, by planting trees of various kinds, among which such graceful conifers as Thujopsis borealis, Thuja gigantea (true), Cupressus macrocarpa, Cryptomeria elegans, etc., should be freely used in the foreground, as in beauty of form they are unsurpassed by any short-lived inhabitants of the summer garden. Except, however, in the case of the Tree-ferns, and various other things not grown in the open air but simply placed there for the summer, it is very desirable not to place the plants in the shade of trees. All the things which have to *grow* in the open air should be placed in the full sun. Not a few hardy subjects will thrive very well without any but ordinary shelter, as, for example, the Yuccas and Acanthuses; but, judging by the remarkable way in which the hardy Bamboo thrives when placed in a sheltered dell, shelter has a considerable influence on the well-being even of these, as it must have on all subjects with large leaf-surfaces. But it should not be forgotten that shelter may be well secured without placing the beds or groups so near trees that they will be robbed, shaded, or otherwise injured by them.

W. R.

March 1, 1871.

PART II.

DESCRIPTION, ARRANGEMENT, CULTURE, ETC., OF SUITABLE SPECIES, HARDY AND TENDER, ALPHABETICALLY ARRANGED.

SUBTROPICAL GARDENING.

PART II.

***Acacia Julibrissin.**—A native of Persia, with large and elegant much-divided leaves, and flowers somewhat like short tinted brushes from the numerous purple stamens. Though this does not succeed as a standard tree in all parts of England (where it grows well against walls, and sometimes flowers), yet doubtless it would do so in some parts of the south, and I have seen it make presentable standards about Geneva and in Anjou. But for our purposes it is better that it should not be perfectly hardy, as by confining it to a single young stem and using young plants, or plants that have been cut down every year, we shall get an erect stem covered with leaves more graceful than a fern, and that is the kind of ornament we want as a graceful object amidst low-growing flowers. The leaves, like those of some other plants of the pea tribe, are slightly sensitive. On fine sunny days they spread out fully and afford a pleasant shade; on dull ones the leaflets fall down. This interesting phenomenon takes place with other members of the same family—for instance, the elegant *A. dealbata* of

* *The names of all hardy species and other kinds easily raised from seed in spring (the kinds useful in all classes of garden), are preceded by an asterisk.*

our conservatories. Seed of *A. Julibrissin*—or the silk-rose, as it is called by the Persians in consequence of its silky stamens—is readily obtained, and it is much better raised from seed, as then you get those single-stemmed and vigorous young plants which are to the flower-garden what an elegant fern is to the conservatory or show-house. To succeed with it in the way above named, it may be protected at the root and cut down every year in spring, or strong young plants may be put out annually, in much the same way as those of *A. lophantha.*

Acacia lophantha.—This elegant plant, though not hardy, is one of those which all may enjoy, from the freedom with which it grows in the open air in summer. It will prove more useful for the flower-garden than it has ever been for the houses, and, being easily raised, is entitled to a place here among the very best. The elegance of its leaves and its quick growth in the open air make it quite a boon to the flower-gardener who wishes to establish graceful verdure amongst the brighter ornaments of his parterre. It has graceful fern-like leaves and a close and erect habit, which permits us to closely associate it with flowering plants without in the least shading them or robbing them. Of course I speak of it in the young and single-stemmed condition, the way in which it should be used. By confining it to a single stem and using it in a young state, you get the fullest size and grace of which the leaves are capable. Allow it to become old and branched, and it may be useful, but by no means so much so as when young and without side branches. It may be raised from seed as easily as a common bedding plant. By sowing it early

ACANTHUS LATIFOLIUS (*lusitanicus*).

Ornamental foliaged herbaceous Section ; retaining its leaves till very late in the year.

in the year it may be had fit for use by the first of June; but plants a year old or so, stiff, strong, and well hardened off for planting out at the end of May, are the best. It would be desirable to raise an annual stock, as it is almost as useful for room-decoration as for the garden. Native of New Holland.

ACANTHUS.

THESE stout and hardy herbaceous plants are of the greatest importance in the subtropical garden or the pleasure-ground, their effect being very good when they are well established. They thrive in almost any soil, but attain their greatest luxuriance and beauty in deep warm ones. The best uses for these species are as isolated tufts in the grass, in the mixed border, or in picturesque groups with other hardy subjects. In all cases they should be placed in positions where they are not likely to be disturbed, as their beauty is not seen until they are well established. All are easily propagated by division. Few herbaceous genera may be made more useful than this.

***Acanthus hirsutus.**—This uncommon species has a narrow spiny leaf, more in the way of *Morina longifolia* than the ordinary Acanthuses, and is dark green in hue. The leaves grow to a length of about 15 ins. or 16 ins. in ordinary soil. Being distinct, it may be worth growing, though in point of character or importance it is inferior to the larger kinds. South of Europe.

***Acanthus latifolius.**—The leaves of this are bold and noble in outline, and the plant has a tendency, rare in some hardy things with otherwise fine qualities, to retain them till the end of the season without losing a particle of their freshness and polished verdure. In fact, the only thing we have to decide about this subject is, what is the best place for it? Now, it is one of those things that will not disgrace any position, and will prove equally at home in the centre of the mixed border, projected in the grass a little from the edge of a choice shrubbery, or in the flower-garden; nobody need fear its displaying anything like the seediness which such things as the Heracleums show at the end of summer. I should not like to advise its being planted in the centre of a flower-bed, or in any other position where it would be disturbed; but in case it were determined to plant permanent groups of fine-leaved hardy plants, then indeed it could be used with great success. Supposing we have an irregular kind of flower-garden or pleasure-ground to deal with (a common case), one of the best things to do with this Acanthus is to plant it in the grass, at some distance from the clumps, and perhaps near a few other things of like character. It is better than any kind of Acanthus hitherto commonly cultivated, though one or two of these are fine. Give it deep good soil, and do not grudge it this attention, because, unlike tender plants, it will not trouble you again for a long time. Nobody seems to know from whence it came. Probably it is a variety of *Acanthus mollis.* The plant varies a good deal; I have seen specimens of it about a foot high, with leaves comparatively small and stiff and rigid, as if cast in a mould, by the side of others of thrice that development, and of the usual texture.

***Acanthus longifolius.**—A fine, distinct, and new species from Dalmatia and S. Europe, 3½ ft. to 4 ft. high, distinguished from *A. mollis* (to which it is allied) by the length and narrowness of its arching leaves. They are about 2½ ft. long, very numerous, of a bright green colour, growing at first erect, then inclining and forming a sheaf-like tuft, which has a very fine effect. The flowers are of a wine-red colour, becoming lighter before they fall. A specimen in the gardens of the Museum at Paris, in four years after planting, had twenty-five blooming-stems rising from the midst of a round mass of verdure nearly 2½ ft. in height and width. This would be very effective on the undulating and picturesque parts of landscape-gardens. It does not run so much at the root as *A. mollis.* It seeds more freely than the other kinds, and may be readily increased by seeds as well as by division. Its free-flowering quality makes this species peculiarly valuable, while it is as good as any for isolation or grouping.

***Acanthus mollis.**—A well-known old border-plant from the south of Europe, about 3 ft. high, with leaves nearly 2 ft. long by 1 ft. broad, heart-shaped in outline, and cut into angular toothed lobes. The flowers are white or lilac, the inflorescence forming a remarkable-looking spike, half the length of the stem. Well adapted for borders, isolation, margins of shrubberies, and semi-wild places, in deep ordinary soil, the richer the better. Increased by division of the roots in winter or early spring.

***Acanthus spinosissimus.**—This is in all respects among the finest of thoroughly hardy "foliage-plants," growing to a height of 3½ ft., and bearing rosy-flesh-coloured flowers in spikes of a foot or more in length. It is perfectly hardy, very free in growth, and is quite distinct from any of the other species, forming roundish masses of dark-green leaves, with rather a profusion of glistening spines, by which it is known immediately from its relatives. As a permanent object, fit to plant in a nook in the pleasure-ground or on the grass, associated with the nobler grasses or other plants, there is nothing to surpass it. I know of no hardy foliage-plant so thoroughly neat in its habit at all times. It does not often flower; and if it should throw up a spike, it will perhaps be no loss to cut it off, as its leaves are its best ornament, though the flowers too are interesting. Never at any time does it require the least attention; it will stand any exposure; and is, in a word, invaluable as a hardy ornamental plant. It will thrive best in good and deep soil. South of Europe.

***Acanthus spinosus.**—This species appears to flower well more regularly than any other. Its leaves are rather narrow, and very deeply divided into almost triangular segments: they are also covered with short spines. The flowering-stems are about 3 ft. high, and bear dense spikes of purplish flowers. Useful for borders, or grouping with the other kinds and plants of similar character and size. South of Europe.

***Adiantum pedatum.**—This fern, which abounds in the woods of Canada and the United States, is unquestionably one of the most elegant of those which are able to endure the climate of Britain, and grows from 16 ins. to 20 ins. high. From the tops of the erect black stems the fronds branch and spread horizontally in a very graceful and peculiar manner. The leaflets are slightly wedge-shaped, the upper margin resembling an arc of a circle. The American Maiden-hair flourishes in a light cool soil, and in a half-shaded position, or in a coarsely-broken, shallow, turfy peat soil, covered with a layer of moss to keep it constantly cool. It is commonly grown in the greenhouse with us, but is especially adapted for embellishing the low and shady parts of rockwork, and for ornamenting beds and mounds of peaty soil which have a north aspect or are sheltered from the full sun. It is propagated by division of the tufts in autumn or early spring. If done in autumn, the divisions should be potted and placed under a frame for the winter, as they form new roots more readily if so treated. There can be no question that, if planted in rich moist soil in a shady wood, we should have no trouble in naturalising this graceful fern, the fronds of which are such graceful objects in the dense woods of the "great country."

Agave americana.—This and its variegated varieties are plants peculiarly suited for subtropical gardening, being useful for placing out of doors in summer in vases, tubs, or pots plunged in the ground, and also for the conservatory in winter. It forms a large rosette of thick fleshy leaves of a glaucous ashy-green colour, overlapping each other at the base, from 4 ft. to 6½ ft. long, and from 6 ins. to 10 ins. broad, ending in a strong spine, and having numerous spines along the margin. When the plant flowers, which it does only once, and after several years' growth, it sends up a flowering-stem from 26 ft. to nearly 40 ft. high. The flowers are of a yellowish-green colour, and are very numerous on the ends of the chandelier-like branches. It will grow in any moderately dry greenhouse or conservatory in winter, or even in a large hall, and may be placed out of doors at the end of May and brought in in October. All the varieties are easily increased from suckers. N. America.

***Agrostis nebulosa.**—This beautiful annual grass forms most delicate feathery tufts about 1 ft. or 15 ins. in height, terminated when in flower by graceful panicles of spikelets, which are at first of a reddish-green colour, and afterwards change to a light red in the upper part, the remaining two-thirds being of a deep green: the pedicels are extremely slender and of a violet colour. It forms very handsome edgings, and is very valuable for bouquets, vases, baskets, room and table decoration, etc. If cut shortly before the seed ripens, and dried in the shade, it will keep for a long time. Dyed in various colours it is much used by makers of artificial flowers. It may be sown either in September or in April or May. In the former case it will flower from May to July, in the latter from July to September. The seed, being very fine, should be only slightly covered. Though small, this deserves a place in groups of the finer and dwarfer plants, such as *Thalictrum minus*, and also in herbaceous borders. Spain.

***Ailantus glandulosa.**—Much trouble and expense are incurred in the purchase, growth, and protection of tender plants with fine compound leaves like this, but which in our climate never display anything like the fresh vigour, health, spotless appearance, and youthful grace characteristic of hardy subjects. This is one of the most valuable of the hardy trees which, if kept in a dwarf state by being planted young and cut down annually, will furnish as good an effect as any tropical plant. The Ailantus should be kept in a young state, with a single stem clothed with its superb pinnate leaves; and we can readily keep it in this form by planting it young and cutting it down annually, taking care to prevent it from breaking into an irregular head, as then the symmetry of the leaf beauty becomes confused and is not at all so effective as when it is kept to a single stem. Vigorous young plants and suckers in good soil will produce handsome, arching, elegantly divided leaves 5 ft. and even 6 ft. long, not to be surpassed by those of any stove-plant. Under such treatment it could be grown conveniently to about from 4 ft. to 7 ft. high, and would thus do grandly for association with the larger class of garden flowers—Gladioli, Dahlias, and Hollyhocks, for example—while among Cannas and the like it will prove fine. The leaves are not liable to be attacked by insects—a good point in a plant used for the purpose I suggest—and they retain their healthy green till the first frosts in November, when they suddenly drop off. It is propagated with facility by cuttings of the roots, but is cheap in all nurseries. China and Japan.

***Aira pulchella.**—One of the most ornamental grasses, with numerous hair-like stems, growing in light elegant tufts 6 ins. to 8 ins. high. It is useful for forming very handsome edgings, or for interspersing amongst plants in borders, or growing in vases or pots for room-decoration.

Its delicate panicles give an additional charm to the finest bouquets. May be sown either in September or in April. S. Europe.

***Alisma Plantago.**—A native perennial water-plant, growing nearly 3 ft. high, and bearing a very handsome pyramidal panicle of rosy-white flowers from June to September. The leaves are oval-lance-shaped with a cordate base, and are borne nearly erect on long stalks for some distance above the surface of the water. A graceful object on the margins of ponds, lakes, etc., where a plant of it transferred from any place where it grows will soon increase.

Alsophila excelsa.—A noble tree-fern, native of Norfolk Island, where it attains a height of 40 ft., crowned with a magnificent circular crest of bipinnate fronds. These fronds or branches fall off every year, leaving an indentation in the trunk. It stands well in the open air in this country in shady, moist, and thoroughly well sheltered places. It should be put out at the end of May, and taken indoors at the end of September or early in October, and receive warm-greenhouse or temperate-house treatment in winter. The same remarks apply to *A. australis*, and probably others of the family will be found to thrive well in the open air when sufficiently plentiful to be tried in that position.

*THE AMARANTUSES.

Among the common annuals of our gardens I know of none more in want of judicious use and appreciation than these. The few we grow are usually treated as rough common annuals, and sown so thickly that they never attain half their true development, or never fulfil any of the graceful uses for which they are adapted. But the family possesses greater claims on our attention by reason of the more recent additions to it. The old "Love lies bleeding" (*A. caudatus*), with its dark-red pendent racemes, is a very striking object when well grown, but *A. speciosus* and some of the more recent varieties are still more so.

***Amarantus caudatus.**—A hardy and vigorous-growing species, from 2 ft. to 3½ ft. high. Flowers from July to September, dark purplish, very small, collected in numerous whorls, which are disposed in drooping spikes so as to form a handsome pendent panicle. There is a variety which has yellow flowers and is equally hardy. It is advisable to give this plant plenty of room to spread; otherwise much of its picturesque effect will be lost; and to use it in positions where its fine and peculiar habit may be seen to advantage,—as, for example, in large vases, edges of large beds of subtropical plants, or dotted among low-growing flowering plants. Although as easily raised as any common annual, it deserves to be properly thinned out, and each plant isolated in rich ground, so that it may attain its full size. E. Indies.

***Amarantus sanguineus.**—Is distinguished by the blood-red colour of its leaves, and grows about 3 ft. high. Its purple flowers appear from July to October, disposed partly in small heads in the axils of the upper leaves, and partly in slender, flexible spikes which form a panicle more or less branching. This plant, though a native of the East Indies, is quite hardy, and seems to do best in light soil with plenty of leaf-mould and having a warm aspect. It may be sown in hotbeds in April and pricked out in May, or in the open air at the end of April or beginning of May, and, like the others, should never be allowed to get crowded.

***Amarantus speciosus.**—A very large kind, well adapted for associating with subtropical plants, as it grows from 3 ft. to nearly 5 ft. high. The flowers are very numerous, of a dark crimson purple, and arranged in large erect spikes, forming a fine plumy panicle. The leaves are suffused with a reddish tinge. Plants of this species are occasionally met with having leaves with a light green centre surrounded by wavy zones of a reddish hue. This colouring disappears at the time of flowering. It is an effective subject in the autumn months. Culture, the same as for the preceding kind. Nepaul.

***Amarantus tricolor.**—Distinguished by the very handsome and remarkable colouring of its leaves, which are of a fine transparent purplish-red, or dark carmine, from the base to the middle. A large spot of lively transparent yellow occupies the greater part of the upper end of the leaf, and sometimes covers it altogether, with the exception of the point, which is mostly green. The leaf-stalk is either of a light green or yellow colour. Sometimes leaves occur which have the lower half green and the upper part red. Another variety (*bicolor*) has leaves of a tender green variously streaked with light yellow. It is rather delicate, and requires very good soil, and a warm, open aspect. Another variety (*bicolor ruber*) is hardier than the last-named, and has leaves which are of a brilliant glistening scarlet when young, gradually changing to a dark violet-red mixed with green. Another variety (*ruber*) has a more squat and ramified habit, and leaves of a deep rose-colour thickly clothing the stems. Other varieties recommended are *elegantissimus* (with scarlet leaves), *Gordoni*, *melancholicus ruber*, and *versicolor*, all having some claims as bedding plants. The foliage of these varieties is exceedingly ornamental, and rivals the finest flowers in the richness of its hues. Planted along with large-leaved subjects, such as the Cannas, Wigandias, Ricinus, Solanums, etc., the effect is very fine. They may also be advantageously employed in borders and flower-beds of all sizes, and for fringing the edges of shrubberies. The varieties of *A. tricolor* are a little more tender than the other kinds, and a light soil and a warmer position are necessary for them. They do well in gardens by the seaside. They should be sown in April in a hotbed, pricked out in a hotbed, and finally planted permanently about the end of May. *A. t. giganteus* is described as very fine in recent catalogues of the nurserymen. To these may be added a beautiful new kind, *A. salicifolius*, in the possession of the Messrs. Veitch, but not yet sent out. It has highly coloured and very long, narrow, and arching leaves, and is a singularly graceful and brilliant object. E. Indies.

***Andropogon squarrosus** is a hardy East Indian grass, which survives the winter with but slight protection, making luxuriant tufts seven feet high, or more, when in flower. It would probably make a beautiful object in the warmer and milder parts of England and Ireland in good soil, but it is not a subject which can with confidence be recommended for every garden. However, all who value fine grasses should try it. Well-drained and deep-sandy loam.

*ARALIAS.

THIS genus embraces many plants of very diverse aspects, and few that are fitted for the open air in our climate; but in the case of *A. canescens*, and its relative (*A. spinosa*), the Angelica-tree of North America, we have subjects which thrive perfectly well in our gardens, and which in the size and beauty of their leaves are far before many "foliage-plants" carefully cultivated in hothouses at a perpetual expense.

***Aralia canescens.**—The specimen of this species figured was one of a batch of young plants growing in a London nursery, and sketched in the summer of 1868. The engraving falls far short of rendering the beauty of the plant. It is easy to imagine what a graceful effect may be realised by such an object, either isolated on the turf near the edge of a shrubbery, or grouped with subjects of similar character. Success with these plants may be secured by first selecting a sheltered and warm position, so that their noble leaves may be well developed and not lacerated by storms when they are fully grown; secondly, by giving them a deep, free, and thoroughly-drained soil; and thirdly, by confining them as a rule to a simple and rather dwarf stem, so that the vigour of the individual may not be wasted in several branches. The effect of a plant kept to a single stem, as shown in the plate, is always much superior to that of a branched one. Young plants present this aspect naturally; but old ones may be cut down,

ARALIA CANESCENS (*japonica*).

Deciduous fine-leaved Shrub ; hardy everywhere.

when they will shoot vigorously. If the effect of a full-grown specimen be desired, the shrubbery is the place for it. = *A. japonica* (*Hort.*).

***Aralia edulis.**—This is a vigorous herbaceous perennial, well suited for those positions in which we desire a luxuriant type of vegetation. It is perfectly hardy, is of a fresh and vigorous habit, and grows 6, 7, and even 8 ft. high in good soil, even so early as the end of June. The leaves attain a length of nearly a yard when the plant is strong, while the shoots droop a little with their weight, and thus it acquires a slightly weeping character. It is rare in this country now, but, being easily propagated, may, it is to be hoped, not long prove so. As it dies down rather early in autumn, it must not be put in important groups, but rather in a position where its disappearance may not be noticed. An isolated position, or one near the margin of an irregular shrubbery, fernery, or rough rockwork by the side of a wood walk, will best suit it. Japan. Division.

***Aralia japonica.**—A valuable species, quite distinct from any of the others, with undivided, fleshy, dark-green leaves. It is usually treated as a green-house plant, but is hardy and makes a very ornamental and distinct-looking shrub on soils with a dry porous bottom. It grows remarkably well in the dwelling-house; in fact it is one of the very few plants of like character that will develop their leaves therein in winter. Not difficult to obtain, it may be used with advantage in the flower-garden or pleasure-ground among medium-sized plants—say those not more than a yard high. It would form striking isolated specimens on the turf, and is also very suitable for grouping. A native of Japan. = *A. Sieboldi.*

Aralia japonica.

***Aralia nudicaulis.**—A very vigorous perennial, with a smooth stem scarcely rising out of the ground, bearing large leaves with long-stalked, oval-oblong, pointed, toothed leaflets, and a shorter naked flower-stem, with from two to seven umbels of blossoms. Roots several feet long and highly aromatic. Similar uses to those directed for *A. edulis.* North America.

Aralia papyrifera (*Chinese Rice-paper Plant*).—This, though a native of the hot island of Formosa, flourishes vigorously with us in the summer months, and is one of the most valuable plants in its way, being useful for the greenhouse in winter and the flower-garden in summer. It is handsome in leaf and free in growth, though to do well it must, like all the large-leaved things, be protected from cutting breezes. If this Aralia be planted in a dwarf and young state, it is likely to give more satisfaction than if planted out when old and tall. The leaves spread widely out near the ground, and then it is very ornamental through the summer. Prefer therefore dwarf stocky plants when planting it in early summer. It should have rich, deep soil and plenty of

Aralia papyrifera.

water during the hot summer months. For the public gardens of Paris it is kept underground in caves during the winter; but in private gardens it will doubtless be thought worthy of a place in the greenhouse throughout that season. In Battersea Park a bed of *A. papyrifera*, 13 ft. in diameter, attained a height of 5 ft., from cuttings struck in the spring of 1868. The plants were left out all the next winter, and, although killed to the ground, the thick fleshy roots next season produced numerous strong shoots or suckers. These were produced irregularly, and so numerously that they had to be thinned out in many places; a few spaces only requiring to be filled up. It is easily increased by cuttings of the root, and is usually planted in masses, edged with a dwarfer plant; but as a small group in the centre of a bed of flowers, or even as an isolated specimen in a like position, it is most excellent.

***Aralia racemosa** (*American Spikenard*).—An herbaceous species, with smooth, widely-branching, diverging stems, about 4 ft. in height, and pinnate, slightly downy leaves with ovate heart-shaped leaflets. Flowers numerous, white, in racemose umbels. Thrives best in good soil in shady or half-shady positions. Similar positions, etc., to those for *A. nudicaulis* and *edulis.* N. America.

***Aralia spinosa** (*Angelica tree of North America*) is highly useful, in consequence of its beauty of foliage, among subtropical plants. Like many of the hardy things, it should not be placed in positions where it would be necessary to remove it, nor closely associated with tender plants requiring frequent disturbance of the soil. Flowers in autumn, small, white, in numerous umbels arranged on a much-branched panicle beset with velvety stellate down. The leaves are twice and thrice pinnate, with ovate, deeply serrated, smooth leaflets. In most cases it grows with a single erect stem—the very type we require—and it should not be allowed to depart from this habit. The stem is fiercely arrayed with spines. On account of its umbrella-like head this fine thing has often been planted in exposed open spaces, where it would produce a distinct feature, but it is better planted where the great leaves will not be lacerated. It generally grows not more than 10 ft. high; and in every size from that down to a plant with a stem not more than 18 ins. high it may be effectively used in the ornamental garden. It is propagated by cuttings of the roots. N. America.

Aralia macrophylla is a fine large digitate-leaved species which stands summer exposure pretty well, but does not make any growth in the open air; hence it can be but of comparatively slight importance for this purpose. Norfolk Island.

Areca sapida.—A New Zealand palm from 6½ ft. to nearly 10 ft. high, with a beautiful crown of bright-green pinnate leaves, which when young are tinged with a bronze colour: leaflets from 16 ins. to 2 ft. in length, lance-shaped. The young leafstalks are of a greyish red hue. A fine palm for placing in the open air in summer, and equally so for the conservatory in winter and spring. It is of very easy culture, if supplied with plenty of water.

Aristolochia Sipho.—This well-known huge-leaved plant is capable of being used with excellent effect where large and distinct foliage is desired. Generally it is used as a wall plant; but it is far finer when used to cover bowers or any like structure. I have seen a most effective object formed by making the framework of a tent loosely with rough boughs, and then planting the Aristolochia around it. It formed a dense green and singular-looking wigwam. *A. tomentosa* is smaller, but distinct in tone of green, well worthy of a place, and to be employed in like manner. N. America.

*__Artemisia anethifolia.__—A hardy perennial species about 4 ft. high, with a simple round stem, woody at the base, and branching vertically above, clothed from about a foot above the ground with much-divided leaves, the segments of which are almost thread-like. Flowers very numerous, small, in a dense, large, terminal panicle, with erect branches. Useful in groups, or as isolated specimens in beds or borders. Division.

*__Artemisia annua__ (*Annual Wormwood*).—An exceedingly graceful kind of wormwood, with tall stems reaching to a height of 5 ft. or 6 ft. in a season; the foliage is small and fine, and the flowers inconspicuous but arranged in not inelegant panicles. The hue of the plant is a peculiarly fresh and pleasing green, and it forms an elegant object in the centre of a flower-bed or group with plants of like character. Raised from seed as easily as any half-hardy annual.

*__Artemisia gracilis.__—An exceedingly graceful plant, 3 or 4 ft. high, with leaves cut into very fine hair-like segments, having some resemblance to fennel or other umbelliferous plants with minutely-cut leaves, and of a deep grass-green, except in the hearts of the shoots, where the young leaves are unfolding, where there is a slight hoary pubescence. The flowers are in compound panicles, quite obscure, of a pale green, not at all ornamental in the common sense, but forming a not ungraceful inflorescence. However, the plant is only likely to be grown for its graceful foliage and habit, and the flowers, which only appear in autumn, may be pinched off if not admired. Similar positions to those recommended for the preceding species. Seed.

***Arum crinitum.**—The appearance of this plant when in flower is very grotesque from the singular shape of its broad, speckled, contorted spathe. The leaves are divided into five or seven deep segments, the central division being much broader than the others, and the leaf-stalks, overlapping each other, form a sort of spurious stem a foot or 14 ins. high, marbled and spotted with purplish-black. The treatment for this plant is similar to that given for *A. Dracunculus;* but as it is rather more tender, it will require more care and shelter in winter. Warm borders, fringes of shrubberies, or beds of the smaller subtropical plants, will suit it best. The appearance of the flower is almost too repulsive for the nerves of some persons. Division. S. of Europe.

***Arum Dracunculus.**—A strange-looking but handsome plant, with a white stem curiously marbled with black, about 3 ft. high, and very deeply cut palm-like leaves, broadly veined and spotted with white. The spathe is of a pale green colour on the outside and of a deep purple-violet within, and, as well as the flowers, exhales a powerful carrion odour. Requires a light, deep, and dry soil, and does best in half-shady positions. Easily multiplied by division in spring or autumn. It is best fitted for the fringes of shrubberies, etc., or among the vegetation that sometimes starts from the bottoms of warm walls. S. of Europe.

***Arum italicum.**—This plant, which is a native of the Isle of Wight and the Channel Islands, resembles the common *A. maculatum* in habit and in the shape of its leaves, which are, however, of a dark bluish-green colour, handsomely spotted with white, and marked with yellow veinings. Although it is a very hardy plant, and will thrive almost anywhere in moist soil and a shady position, it will be better to place it in sheltered positions along the sunny fronts of shrubberies, amidst low-spreading evergreens, and in cosy spots about the flanks of rockwork and ferneries, to prevent its handsome foliage from being disfigured by cold wintry winds. One great merit of this is that it may be used to ornament positions in which few other plants will thrive,—as, for instance, under trees, groups of shrubs, etc. Easily multiplied by division in the end of summer and in autumn. S. of Europe.

***Arundo conspicua.**—A companion for the Pampas grass, though by no means equal to it, as has been stated by some writers. In some very fine deep and free sandy loams it attains a height of nearly 12 ft., but this is rare. As a conservatory subject it is fine in flower, and it will be often seen in large conservatories after a few years. A large pot or tub will be necessary if it is grown indoors. The drooping foliage will always prove graceful, and it sends up long silvery plumes, drooping also, and strikingly beautiful. Judging by its different appearance when freely grown in a tub indoors, and when planted out even in favourable spots, my impression is that it by no means takes so kindly to our northern climate as the Pampas grass. However, it is well worth growing, even in districts where it does not attain a great development. It comes into flower before the Pampas grass, and may be considered as a sort of forerunner of that magnificent herb. New Zealand.

***Arundo Donax.**—This great reed of the south of Europe is a very noble plant on good soils. In the south of England it forms canes 10 ft. high, and has a very distinct and striking aspect. It will grow higher than that if put in a rich deep soil in a favoured locality; and those who so plant clumps of it on the turf in their pleasure-grounds will not be disappointed at the result. Nothing can be finer than the aspect of this plant when allowed to spread out into a mass on the turf of the flower-garden or pleasure-ground. It seems much to prefer deep sandy soils to heavy ones; indeed, I have known it refuse to grow on heavy clay soil, and flourish most luxuriantly on a deep sandy loam in the same district. Like all large-leaved plants, it loves shelter. No garden or pleasure-ground in the southern parts of England and Ireland should be without a tuft of it in a sheltered spot. But, fine as it is for effect and distinctness, its variegated variety is of more value for the flower-garden proper.

***Arundo Donax versicolor.**—We have already noticed several fine things for grouping together, or for standing alone on the turf and near the margin of a shrubbery-border, and this is as well suited for close association with the choicest bedding-flowers as an Adiantum frond is with a bouquet. It will be found hardy in the southern counties, and, considerably north of London, may be saved by a little mound of cocoa-fibre, sifted coal-ashes, or any like material that may be at hand. In consequence of its effective variegation, it never assumes a large development, like the green or normal form of the species, but keeps dwarf, and yet thoroughly graceful. It is of course best suited for warm, free, and good soils, and abhors clay, though it is quite possible to grow it even on that with a little attention to the preparation of the ground. But it is in all cases better to avoid things that will not grow freely and gracefully on whatever soil we may have to deal with: and it is to those having gardens on good sandy soils, and in the warmer parts of England, that I would specially recommend this grand variegated subject. For a centre to a circular bed nothing can surpass it in the summer and autumn flower-garden, while numerous other charming uses may be made of it. Not the least happy of these would be to plant a tuft of it on the green turf, in a warm spot, near a group of choice shrubs, to help, with many other things named, to fill up the gap that is now nearly everywhere observed between ordinary fleeting flowers and the taller tree and shrub vegetation. It is better to leave the plant in the ground, in a permanent position, than to take it up annually. Protect the roots in the winter, whether it be planted in the middle of a flower-bed or by itself in a little circle on the grass. Increased by placing a shoot or stem in a tank of water, when little plants with roots will soon start from every joint; they should be cut off, potted, and placed in frames, where they will soon become strong enough for planting out.

***Arundo Phragmites** (*Common Reed*).—A native marsh- or water-plant, 5 ft. or 6 ft. high, bearing when in flower a large, handsome, spreading, purplish panicle. The stems are smooth, simple, very erect, and grow closely together. The plant is only attractive when in flower, as its flat, ribbon-like leaves do not of themselves present any very striking appearance. Useful for the margins of artificial waters, etc., to which it may be brought from its wild haunts. It should, however, if possible, be kept in one spot and not allowed to spread too much.

***Asclepias Cornuti.**—A handsome hardy perennial from N. America, sending up from its running, underground rootstock a number of erect, unbranched stems, from 4 ft. to 6½ ft. high, thickly covered with large opposite oval leaves. The flowers are of a light rose-colour, and agreeably fragrant, and are borne in large umbels at the tops of the stems. The plant does well in almost any kind of soil or position, and is well adapted for planting in places which do not require much attention. As it spreads very rapidly at the root, it is better to exclude it from small beds or gardens, which would soon be overrun by it. Several other species are in cultivation, the best of which are *A. speciosa* (Douglasii), *A. incarnata*, and *A. tuberosa*.

***Asparagus Broussoneti** (*Giant Asparagus*).—A vigorous, climbing asparagus, with a tapering, shrubby stem, 10 or 12 ft. high. The flowers are small and inconspicuous, and are followed by numerous small red berries. An excellent subject for tall trellises, rustic bowers, stumps of trees, tall poles, etc. Canary Islands.

Asplenium Nidus-avis.—This is a remarkable fern, which has been placed out of doors in the garden in summer, from early in June to October; but it is not vigorous or hardy enough to be generally recommended for this purpose. However, as it may have been noticed in abundance at Battersea Park, I allude to it here. The leaves are rather broad, pointed, and undulating, nearly 3 ft. long, and form roundish, spreading, nest-like tufts. It is a favourite subject in places where large collections of tropical ferns are grown, and in such places a plant may be tried in the open air in a very warm, shady, and perfectly sheltered position. E. Indies.

Asplenium Nidus-avis.

***Astilbe rivularis.**—A large-leaved and striking plant from Nepaul, with the habit and general appearance of a *Spiræa*, growing to a height of more than 3 ft., and of a free and graceful habit, which makes it useful for association with the finer-foliaged herbaceous plants, and for dotting here and there in the wild or picturesque garden. It keeps its foliage well through the season, unlike some herbaceous plants, and is therefore all the more valuable. Flowers late in summer, small, yellowish-white, in large panicled spikes. The radical leaves are broad, twice ternate with toothed divisions, and the base of the leaf-stalk is covered with numerous rough tawny hairs. Being pretty hardy, the Astilbe usually succeeds well in any cool rich soil, and best in half-shaded positions. Easily multiplied by division. It is suited for isolation, borders, fringes of shrubberies, or for groups of hardy plants.

***Astilbe rubra.**—A very pretty and hardy plant, also resembling a *Spiræa* in habit and appearance, and growing from 4 ft. to 6 ft. high. The leaves are twice ternate, with oblique, heart-shaped leaflets, 1 in. to 2 ins. long, and with lengthened serrated points. The flowers are numerous, in dense panicles, and of a rose-colour, appearing late in summer and in autumn. The same positions, etc., as those for the preceding kind. North America, Japan, and mountains of Northern India.

***Bambusa.**—I wish to call the attention of all horticulturists who live in the southern and more favoured parts of these islands to the fact that there are several bamboos and bamboo-like plants from rather cool countries that are well worth planting. Nothing can exceed the grace of a bamboo of any kind if freely grown; but if starved in a crowded house, or grown in a cold dry place, where the graceful shoots cannot arch forth in all their native beauty, nothing can be more miserable in aspect. In cold bad soils and exposed dry places in the British Isles these bamboos have little chance; but, on the other hand, they will be found to make most graceful objects in many a sheltered nook in the south and south-western parts of England and Ireland. We have some knowledge of the capabilities of one kind in this country. In a well-sheltered moist spot at Bicton I have seen *Bambusa falcata* send up young shoots, long and graceful, like the slenderest of fishing-rods, while the older ones were branched into a beautiful mass of light foliage of a distinct type. The same plant grows in the county of Cork to a height of nearly 20 ft. This is the best known kind we have. At Paris I was fortunate enough to observe various other kinds doing very well indeed, although the climate is not so suitable as that of Cork or Devon. These were *Bambusa edulis*, *aurea*, *nigra*, *Simonii*, *mitis*, *Metake*, and *viridi-glaucescens*, the first and last of this group being very free and good. All the others will prove hardy in the south of England and Ireland, though, as some of them have not yet been tried there, it requires the test of actual experiment. Those who wish to begin cautiously had better take *B. Simonii*, *viridi-glaucescens*, and *edulis* to commence with, as they are the most certainly hardy, so far as I have observed. The best way to treat any of these plants, obtained in summer or autumn, would be to grow them in a cool frame or pit till the end of April, then harden them off for a fortnight or so, and plant out in a nice warm spot, sheltered also, with good free soil—taking care that the roots are carefully spread out, and giving a good free watering to settle the soil. There are no plants more worthy of attention than these where the climate is at all favourable, and there are numerous moist nooks near the sea-side where they will be found to grow most satisfactorily, as well as in the south.

***Bambusa aurea.**—A very hardy and graceful Chinese species, differing but slightly from *B. viridi-glaucescens* in size and habit, and forming elegant tufts with its slender much-branched stems, which attain a height of from 6½ ft. to 10 ft., and are of a light-green colour when

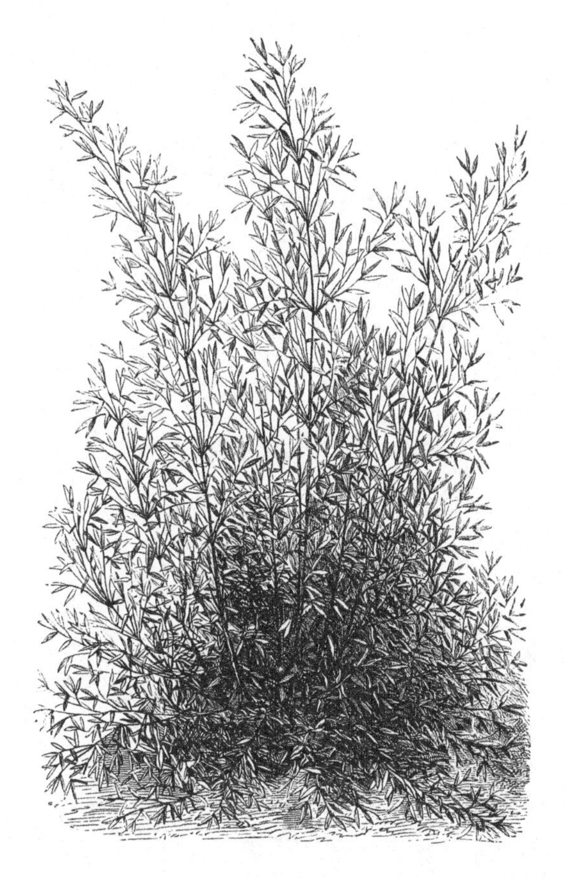

BAMBUSA AUREA.

Thriving in warm and mild southern districts.

young, changing into a yellowish hue, and finally becoming of a straw-yellow when fully grown. The leaves are lance-shaped acute, light green, and are distinguished from those of *B. viridi-glaucescens* by having their under surface less glaucescent, and the sheath always devoid of the long silky hairs. The preliminary remarks on culture, etc., will apply to all the species here described.

***Bambusa edulis.**—A hardy and vigorous kind, with very elegant light-green shoots and olive-green stems, attaining a height of 10 ft. in the neighbourhood of Paris. The leaves are small, and the plant is not nearly so branching as in some other kinds.

***Bambusa falcata** (*Arundinaria falcata*).—A very ornamental species from Nepaul and the Himalayas, and at present the only kind of bamboo much planted with us. It grows from 7 ft. to 20 ft. high, and has woody, twisted, smooth stems of a yellowish-green or straw-colour, knotty, bearing on one side of each of the knots a bundle of small branches equally knotty and twisted. The whole plant has a pale yellowish hue, except in the young spikelets and sheaths, which are occasionally purplish. The leaves are of a fine delicate green, from 4 ins. to 6 ins. long, ribbon-like, linear-acute, sickle-shaped, in two rows, short-stalked, and sheathing. It is hardy over the greater part of England and Ireland, but only attains full development in the south and west. I have seen it attain great luxuriance in Devon, and nearly 20 ft. high near Cork, though in many districts it is stunted. It loves a deep, sandy, and rich soil, and plenty of moisture when growing fast.

***Bambusa Fortunei.**—A pretty dwarf variegated

species from Japan, of which I have not seen the green form in cultivation, with very dwarf, slender, branching, hollow, half-shrubby stems, seldom growing more than 18 ins. high, and with very short internodes. The leaves are 3½ ins. to 8 ins. long, linear-lance-shaped, abruptly pointed, somewhat rounded at the base, serrated and often fringed with long hairs on the margin, downy on both sides; they are distinctly variegated, the transverse veins being often of a bottle-green colour; stalks very short and hairy. This kind has proved hardy in our gardens, but it has not the charm of grace possessed by the other kinds, and is chiefly desirable in collections of variegated and edging plants.

***Bambusa japonica** (*B. Metake*).—A large-leaved and rather dwarf species from Japan, growing from 4 ft. to 7 ft. high, with erect thickly-tufted stems, which are entirely covered by the sheaths of the leaves; the branches are also erect. The leaves are lance-shaped, with a very sharp point, dark green, persistent, narrowed into a short leaf-stalk, and nearly a foot long. This species sometimes flowers with extraordinary profusion at the expense of a portion of the foliage, which withers away and leaves the naked stems exposed. This may, however, be prevented to some extent, by placing the plants on mounds somewhat above the level of the surrounding soil. I have seen it thrive very freely in the late Mr. Borrer's garden in Sussex, and in one or two other places. It loves a peat soil, or a very free moist and deep loam, and runs a good deal at the root.

Bambusa mitis.—A fine and vigorous kind from Cochin China, somewhat tenderer than most of the other

BAMBUSA FALCATA.

Hardy Bamboo Section; growing 16 to 20 feet high in the milder southern parts of England and Ireland.

kinds enumerated, though no doubt it will be found to thrive in the milder southern districts; or it may be found useful if grown in the conservatory in winter and placed out in the open air in summer, as is sometimes done with *B. arundinacea*, which otherwise could not possibly be seen out of doors in our climate. Panicle simple, erect, close; spikes long, imbricated. Leaves rather large, lance-shaped, striated, clasping the stem, which is woody and tapering; nodes rather distant, and not very prominent.

***Bambusa nigra.**—A rather compact-growing Chinese kind, with nearly solid stems, and thinner leaves than those of any other species. The stems are smooth and bushy, about 7 ft. high, of a light green, dotted and striped with purple when young, changing to a glistening black when fully grown; they branch very much at the top, and sometimes from the base up. The leaves are oval-oblong, acute, shortly-stalked, with a hard, dry, persistent sheath; their tender green colour contrasting finely with the blackish hue of the stems. It is best planted as isolated specimens near the margins of shrubberies, or on slopes in the pleasure-ground in warm, sunny, and sheltered positions, in deep, sandy, and well-drained soil.

***Bambusa Quilioi.**—A Japanese species of vigorous growth, with robust green stems and bright-green leaves, polished above and slightly mealy beneath, the ligule bearing a little bundle of brownish-grey hairs at the top. This kind I first saw in the gardens of the Acclimatisation Society at Paris, where it was thriving vigorously, and I have little doubt of its proving valuable in Britain.

***Bambusa Simonii.**—A handsome, distinct, and

vigorous species, which has grown very freely for some years past in the neighbourhood of Paris. The stems are numerous and grow as much as 10 ft. high in a season. They are mealy-glaucous at the joints, and the branchlets are numerous and rather closely crowded. The leaves are narrow, sometimes nearly a foot long, and are occasionally striped with white. This species, which was obtained from China some years since, has thriven very well in the gardens at Paris, where M. Carrière first drew my attention to it. From what I have seen it do there I have no doubt it will prove of great value in the milder southern parts of England and Ireland.

***Bambusa violascens.**—A hardy and vigorous kind, intermediate between *B. nigra* and *B. viridi-glaucescens*, most resembling the last-mentioned however. It has blackish-violet much-branched stems, which assume a yellow tinge with age. The leaves are green above, bluish-grey beneath, with an elongated ligule surrounded by a bundle of black hairs. Native of China.

***Bambusa viridi-glaucescens.**—A species from Northern China, which has been proved very hardy and free in the Paris gardens, and will, probably, in warm parts of our islands, make a more vigorous growth and prove a more beautiful object than any other kind. The stems, of a light-yellowish-green, grow from 7 ft. to 12 ft. high, branching from the base, each branch again branching very much. The leaves are very numerous, especially at the ends of the branches, of a pale-green, bluish underneath, sheathing the stem for a considerable length. It forms a fine object when planted as isolated specimens in sheltered warm glades in the pleasure-ground, or in snug open spots near wood-walks, in very deep, rich, light, and well-drained soil.

***Bambusa viridis striata.**—Described as a vigorous-growing, hardy kind, with numerous branches, having its long leaves green on both sides, and marked with bands, some of a yellowish and others of a deeper green. It is a native of Japan, and was recommended by MM. Thibaut and Ketteleer of Paris, but I have had no experience of its growth.

***Baptisia australis.**—A handsome hardy perennial from N. America, forming strong bushy tufts from 3 ft. to 5 ft. high, and from 3 ft. to 6 ft. across, with sea-green trifoliate leaves which reflect a metallic lustre. The flowers are of a delicate blue, with wings of a greenish-white colour, and are borne in long erect spikes. Grows well in ordinary, deep, well-drained soil, preferring a sandy loam. *B. exaltata* and *B. alba* are closely allied to the preceding species, and form equally handsome bushes. The foliage of these is of a character to permit of their being grouped along with some of the finer perennial foliage-plants with good effect. Division.

***Berberis Aquifolium.**—A well-known shrub from N. America, with simple stems from 3½ ft. to 6½ ft. high, and very glistening, bright green leaves, each consisting of 7 or 9 sessile, oval, spiny leaflets. Where the plant is fully exposed to the sun, the foliage frequently acquires an agreeable reddish tinge. There are numerous varieties, of which *B. floribunda* may be mentioned as very handsome in habit and profuse in flower, and *B. nitens*, remarkable for the extremely glossy appearance of the old leaves, which when young are of a rich bronze changing to a dark green. This variety is of a dwarf and compact habit. It requires a shady situation and a compost of peat, loam, and sand. Though so very common, it will be found worth planting in some places among groups of hardy things, and also for isolation on the turf, its leaves being very ornamental.

***Berberis Bealii.**—This is perhaps the finest of all the hardy species, whether as regards foliage or flower, while the fruit, in colour and size, surpasses that of any other kind. The leaflets vary to a very great degree on the same plant, both in form and size, some being 5 ins. long and 3½ ins. broad, the average size being 3 ins. long by 2 ins. broad: some again are nearly square, while others are long and narrow, with a very stiff triangular point. Shade, shelter from gusts of wind, and rich, well-manured soil are absolutely necessary to do full justice to the merits of this species. By pruning it to a single stem, it may be made to assume a very effective palm-like character. It is easily propagated from seed; a single berry frequently producing three plants. Being a noble subject for quiet half-shady wood-walks in peat or moist sandy loam, it should be much planted in the southern and milder districts. Where it thrives freely, it would prove a fine object on the margins of shrubberies grouped with the hardy "subtropical" plants, or indeed in any position.

***Berberis japonica.**—A very handsome species, 5 ft. or 6 ft. high, with very leathery, pinnate, spiny leaves, from 1 ft. to 18 ins. long, slightly tinged with pink when first opened, then becoming pale green, and finally changing to dark green blotched with yellow. Flowers yellow, in large racemes, succeeded by large handsome

BERBERIS NEPALENSIS

Fine-leaved evergreen shrub Section; very effective when well-grown, in the warmer southern districts.

clusters of purple berries. This is also a fine kind for half-shady walks, and for similar positions and soil to those recommended for *B. Bealii.* Japan.

***Berberis nepalensis.**—The noble habit of this plant makes it peculiarly valuable, possessing, as it does, the grace of a luxuriant fern with the rigidity of texture and port of a Cycas. The leaves are occasionally 2 ft. in length and of a pale green colour, sometimes with eight pairs of leaflets and an odd one: some of the leaflets 6 ins. long and nearly 2 ins. broad, with coarse spiny teeth on the margin. The inflorescence is very striking and beautiful. The Nepaul Barberry is one of those subjects that are too hardy to perish in our climate, yet which do not usually attain perfect development in it. It exists about London in the open air, and flowers in early spring; but the leaves seldom attain one-fourth of their full development, and the plant scarcely ever displays its vigorous grace. In mild parts, principally in the south and south-west, it grows more freely, and when judiciously placed in sheltered positions, in deep and rather sandy soil, it becomes a beautiful object. Where it thrives in the open air, it may be most tastefully used in the more open spots near the hardy fernery, here and there among "American plants," or other choice shrubs with simple leaves, and also isolated in the grass a little way from the margin of the shrubbery in sheltered spots in the pleasure-ground. It should also, in places favourable to its growth in the open air, prove very useful as a hardy "subtropical" plant. Where it does not thrive well in the open air, it should not be planted. Nepaul.

***Beta cicla variegata** (*Chilian Beet*).—Under this name a very showy plant has recently come into cultivation. When well grown the leaves are often more than a yard long, and present a vivid and most striking coloration. Their midribs are 4 ins. or more across, and vary from a dark deep waxy orange to vivid polished crimson. The splendid hue of the lower part of the leaf-stalk flows on towards the point, and spreads in smaller streams through the main veins and ramifications of the great soft blade of the leaf, which is often 1 ft. and even 15 ins. in diameter, if the plant be in rich ground. The under sides of the leaves are most richly coloured, and the habit such that these sides are well seen. It requires the treatment of an annual—to be raised in a gently heated frame, and afterwards planted out in very rich ground, though it may also be kept over the winter in pots. It varies a good deal from seed, and the most striking individuals should be selected before the plants are put out. Used sparingly, its effect would perhaps be more telling than if in quantity, and it is well suited for isolation. Chili.

***Bocconia cordata.**—This is a fine plant in free soil, but comparatively poor in that which is bad or very stiff. It forms handsome erect tufts from 5 ft. to over 8 ft. high, and is admirably suited for embellishing the irregular or sloping parts of pleasure-grounds. The stems grow rather closely together, and are thickly set with large, reflexed, deeply-veined, oval-cordate leaves, the margins of which are somewhat lobed or sinuated. The flowers, which are rosy-white and very numerous, are borne in very large terminal panicles. The flowers are not in themselves pretty, but the inflorescence, when the plant is well grown,

BLECHNUM BRASILIENSE

Dwarf tender Tree Fern : in sheltered shady dells during the summer months.

has a distinct and pleasing appearance. The plant is seen to best effect when isolated, and does well in ordinary garden-soil or free sandy loam. It attains its greatest size when placed in the angle of two walls which shelter it from the north and east, which seems to indicate that it does not like sudden changes of temperature and light. It should not be stirred too often, nor divided for several years. It will prove a good thing for associating with other fine hardy plants in bold groups. Seed or cuttings. China.

Bocconia cordata.

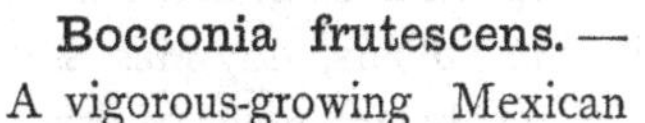

Bocconia frutescens. — A vigorous-growing Mexican shrub, 3½ ft. to nearly 6 ft. high, with few and very brittle branches, large, sea-green, handsome leaves, and greenish flowers. Very effective when placed on grass-plats, either in groups, or as isolated specimens. It requires a somewhat warmer climate than ours to thrive well, though it is sometimes seen in fair condition in the London parks. A mixture of free sandy loam and peat, well-drained ground, and an airy position are necessary. Multiplied by sowing in a hotbed in spring, and may be placed out from June to the end of September. It is difficult to propagate it by cuttings.

***Brassica oleracea crispa.**—A handsome kind of cabbage nearly 4 ft. high, with elegantly-cut arching leaves, the divisions of which are finely curled or frizzled. In autumn and winter it may be advantageously employed in the embellishment of winter-gardens, the leaves being at their best during that part of the year.

A still more striking subject is *B. o. palmifolia*, which attains a height of 6½ ft., and bears its leaves near the summit of the stem, having quite a palm-like appearance in the end of the summer and in autumn. This kind might be used with good effect in various positions, as its "cabbage" character is not so evident. The fact of their being cabbages prevents many people from using these really ornamental plants.

The variegated Kales have been much employed and with a very good result in the winter-garden at Wardie Lodge in Scotland and in other places: they lose their beauty early in spring. "To keep them dwarf and compact, and to bring out their fine colours," say Messrs. Stuart and Mein, "we sow at the beginning of March thinly. After the plants are a moderate size, we transplant them into a poorish soil, in an open space of ground, but not too closely together. They remain there until they are wanted for use, when they are removed at any time to the winter-garden. We plant out into our beds in November, and keep the heads with their foliage close to the ground. We cut off all the lower rough leaves, leaving the rich-coloured head or centre, which in all weathers will be found to stand up neat and trim, even in bright frosty weather. As the plants are thus much reduced in size, they can be planted much closer in the beds. We may add that, as in other hybrids, worthless plants will occasionally make their appearance: these we discard. No plants should be selected for the beds except those showing pleasing shades of colour."

Brexia madagascariensis. — A handsome shrub with a slender erect stem (which sometimes attains a height of 25 ft. or 30 ft. in its native country, but with us is seldom seen more than a fourth of that), clothed with alternate, leathery, long, rather narrow, light-green leaves, nearly or quite smooth at the margin. It is one of the tropical stove-plants that have stood well in the open air from June to early in October, but very few places can spare it for this purpose. It requires ordinary stove culture during winter and spring, and should only be placed out after having made a strong growth, and having that growth hardened off. Madagascar.

***Buphthalmum speciosum.** — A hardy, distinct, and vigorous herbaceous plant, the stems of which are stout, very slightly branching, and about 4 ft. high, with broad, oval-acute leaves mostly clustered around the base of the plant, the lower ones falling gracefully towards the earth. The flowers, which have a red or purple disk and yellow rays, are more than 2 ins. across, and are terminal, solitary, long-stalked, borne in the axils of the upper leaves, and appear in June, July, or August, according to the season. The plant seldom

Buphthalmum speciosum.

flowers well before the third year. It is of easy culture in any soil, is increased by division in autumn, winter, or spring, and is best fitted for association with the more vigorous herbaceous plants in rough places. S. of Europe.

Caladium esculentum.—This species has, for outdoor work, proved the best of a large genus with very fine foliage. It is only in the midland and southern counties of Great Britain that it can be advantageously grown, so far as I have observed; but its grand outlines and aspect when well developed make it worthy of all attention, and of a prominent position wherever the climate is warm enough for its growth. It may be used with great effect in association with many fine foliage-plants; but *Ferdinanda*, *Ricinus*, and *Wigandia* usually grow too strong for it, and, if planted too close, injure it. This may have been noticed particularly in cases where it was used as a bordering to masses of the strong-growing kinds above named. For all kinds of stonework, vases, etc., it is peculiarly effective and beautiful. This plant, requires, above all others, a thoroughly-drained, light, rich, warm soil. In times of great heat, it should be plentifully watered, and occasionally with liquid manure. The month of May is the best time for planting it out; and if groups are formed, the plants should have an interval of 2 ft. or 2½ ft. between them. The foliage generally arrives at its full beauty and development in August and September. At the approach of cold frosty weather, all the leaves, or all but the central one, should be cut down to within an inch or two from the crown, and a few days afterwards the tubers should be taken up and left on the ground for a few hours to dry:

CALADIUM ESCULENTUM.

Tender Section; displaying noble leaves during summer in the warmer parts of the southern counties.

COLOCASIA ODORATA.

Tender stove Section; will endure exposure only during summer in the warmest parts of the southern counties.

they should then be stored on the shelves of a greenhouse, or in a cellar or other place where they will be sheltered from frost and moisture. By placing the tubers in a hotbed in March, plants may be obtained with well-grown leaves for planting out in the open air about the end of May or the beginning of June. New Zealand.

Caladium odorum (*Colocasia odora*).—A very striking plant, with stout stems usually from 3 ft. to 8 ft. in height, but growing much taller in a warm stove. The leaves are erect, very broad, and heart-shaped, marked with strong veinings, and frequently measure more than 3½ ft. in length. The flowers are exceedingly fragrant. It is a fine subject for isolation on grass-plats, its tall arborescent habit distinguishing it from all the other species; but it is unfortunately too tender to thrive in our climate except in sunny sheltered dells in the southern parts, and should not be planted out until June. E. Indies.

***Calla æthiopica** (*Lily of the Nile*).—This well-known plant may be grown either as an aquatic in pieces of ornamental water, fountain-basins, etc., or in the open ground in cool, moist soil, and equally well in positions exposed to the full sun and in those which are shaded. Being so very distinct in leaf and beautiful in bloom, this old favourite will be seen to as much advantage grouped with the smaller fine-leaved plants in beds as ever it has been in our stoves or windows. S. Africa.

*THE CANNAS.

IF there were no plants of handsome habit and graceful leaf available for the improvement of our flower-gardens but these, we need not despair, for they possess almost every quality the most fastidious could desire, and present a useful and charming variety. The larger kinds make grand masses, while all may be associated intimately with flowering-plants—an advantage that does not belong to some free-growing things like the Castor-oil plant. The Canna ascends as boldly, and spreads forth as fine a mass of leaves as these, but may be closely grouped with much smaller subjects. The general tendency of most of our flower-garden plants is to assume a flatness and dead level, so to speak; and it is the special quality possessed by the Cannas for counteracting this that makes them so valuable. Even the grandest of the other subjects preserve this tameness of upper-surface outline when grown in great quantities: not so these, the leaves of which, even when grown in dense groups, always carry the eye up pleasantly from the humbler plants, and are grand aids in effecting that harmony which is so much wanted between the important tree and shrub embellishments of our gardens and their surroundings, and the dwarf flower-bed vegetation. Another good quality of these most useful subjects is their power of withstanding the cold and storms of autumn. They do so better than many of our hardy shrubs and plants, so that when the last leaves have been blown from the Lime, and the Dahlia and Heliotrope have been hurt by frost, you may see them waving as gracefully and as green as the vegetation of a temperate stove. Many of the subtropical plants, used for the beauty of their leaves, are so tender that they go off in autumn, or require all sorts of awkward protection at that season; but the Cannas last

CANNA.

The most important and generally useful of tender plants for our climate.
Many kinds are hardy if protected in winter.

in good trim till the borders must be cleared. All sheltered situations, places near warm walls, and nice snugly-warmed dells, are suitable positions for them. They are generally used in huge and ugly masses, both about Paris and London; but their true beauty will never be seen till we learn to place them tastefully here and there among the flowering-plants—just as we place sprigs of graceful fern in a bouquet. A bed or two solely devoted to them will occasionally prove very effective; but enormous meaningless masses of them, containing perhaps several hundred plants of one variety, are things to avoid and not to imitate. As to culture and propagation, nothing can be more simple: they may be stored in winter, as readily as potatoes, under shelves in the houses, in the root-room, or, in fact, anywhere if covered up to protect them from frost. And then in spring, when we desire to propagate them, nothing is easier than pulling the roots in pieces, and potting them separately. Afterwards it is usual to bring them on in heat, and finally harden them off previous to planting out in the middle of May; but a modification of this practice is desirable, as some kinds are of a remarkably hardy constitution, and make a beautiful growth if put out without so much as a leaf on them. The soil for all Cannas should be deep, rich, and light.

In rambling through the suburbs of Paris, I once came upon a tuft of Canna springing up strongly through a box-edging—pretty good evidence that it had remained there for some years. Upon inquiring of the proprietor of the garden I found this was the case, and that he had no doubt of the hardiness of several other kinds. They were planted not more than 8 ins. or 10 ins. deep. When we remember that the Cannas are amongst the most valuable plants we use for giving grace and verdure to the flower-garden, this surely is a hint worthy of being acted upon, as, of course, they will prove equally hardy with us. Considering their diversity of colour and size, their graceful pointed habit and facility of propagation, we must concede them the first place; but their capability of being used by anybody who grows ordinary bedding-plants, and the fact that they may be preserved so very easily through the winter, enhance their value still more. Cannas, protected by a coating of litter, have been left out in Battersea Park through severe winters, and during the unfavourable summer of 1867 attained a height of nearly 12 ft. Where it is desired to change the arrangements as much as possible every year, it may not be any advantage to leave them in the ground, and in that case they may be taken up with the bedding-plants, and stored as simply and easily as carrots. Wherever they are grown as isolated tufts, in small groups, or in small beds, it will be best not to take them up oftener than every second or third year. These noble plants would also adorn the conservatory, which is often as devoid of any dignified vegetation as the unhappy flower-gardens which are seen all over the country. Few subjects would be more effective, none more easily obtained.

SPECIES AND VARIETIES OF CANNA.

In the following list of the species and varieties of Canna, the first thirteen kinds are considered species:

but the finest kinds for garden use will be found among the Hybrids and Varieties.

Canna aurantiaca.—A vigorous kind, 6½ ft. or more in height, with large, broadly oval-lance-shaped leaves, of a pale green colour, slightly waved on the margin. Flowers with rose-coloured outer, and reddish inner, divisions, the upper lip being of an orange colour and the lower one yellow dotted with orange. Rhizome tuberous, with long subterranean shoots. Antilles.

Canna discolor. — One of the finest kinds, and, although it does not flower regularly in the open air, one of the most valuable on account of its foliage. The leaves are very large, broadly oval-oblong, the lower ones tinged with a blood-red hue, and the upper ones veined or streaked with purple. Stems reddish, stout, growing to a height of more than 6 ft. Flowers orange-red, with bright-red inner divisions. S. America.

Canna edulis.—A vigorous species, largely cultivated by the Peruvians for the sake of its edible roots, growing from 6 ft. to 7 ft. high, the stems tinged with deep purple. Leaves broadly oval - lance - shaped, green tinged with purple maroon. Flowers large : external divisions purple, upper internal division scarlet shaded with yellow, the lower one of an orange-red. S. America.

Canna flaccida.—A species remarkable for the great size of its flowers, which bear some resemblance to those of *Iris Pseud-acorus.* They are entirely yellow, flaccid, few in number, and very transitory. This is not a tall species, seldom exceeding 2 ft. 8 ins. in height. Leaves oval-lance-shaped, erect, glaucescent. South America. This species should not be confounded with the *C. flaccida* of Willdenow which is found on the shores of the Mississipi.

Canna Gaboniensis.—A species from the Gaboon River, with the habit of *C. nepalensis.* Rhizomes large and round; stalks green, large; leaves deep green lightly edged with purple, longish, acuminate; flowers medium-sized, clear orange; habit fine, special; height 5½ ft. to 6 ft.

Canna gigantea.—A robust kind, growing about 6½ ft. high. Flowers in summer, large and very handsome; external divisions orange red; internal ones deep purple-red; spathes reddish. Leaves more than 2 ft. long; stalks covered with a velvety down. S. America.

Canna indica.—Flowers in summer, tolerably large, irregular, in erect spikes; external divisions light yellow; upper inner divisions of a carmine-red; the lower one yellow dotted with carmine. Leaves large, alternate, oval lance-shaped; the stalks sheathing at the base. Stems 3 ft. to 6 ft. high. A variety (*C. i. superba*) has much larger flowers of a scarlet colour. India. It is not nearly so useful or imposing in appearance as some of the newer hybrid kinds.

Canna iridiflora. — Flowers in midsummer, large, lively rose-colour, with a yellow spot on the lip; divisions of the calyx red, erect, oval-acute; spikes somewhat drooping, issuing several together from the same spathe. Leaves broadly oval-acuminate, slightly membranous at the margin, and having some hairs at the lower part of the midrib. Stems 6 ft. to over 8 ft. high. Peru. A somewhat tender species, and although one of the finest, does not flower freely. To secure its blooming, the plants should be taken up towards the end of summer, potted, and kept in a hothouse through the winter. They should be watered moderately, and, treated in this way, will almost always begin to show flower in the ensuing spring.

Canna limbata.—This has numerous stems nearly 3 ft. high, and oblong-lance-shaped, acute leaves. Flowers in long loose spikes issuing from glaucous spathes tinged with red; outer divisions pale yellow; upper lip scarlet edged with a yellow margin; lower lip red, with golden reticulations. Brazil.

Canna musæfolia. — This species was formerly described in the English, Dutch, and German horticultural journals under the name of *C. excelsa.* It was named *musæfolia* by M. Année, who introduced it into France in 1858, from the resemblance of its leaves to those of the Musa or banana-tree. It reaches a height of more than 8 ft., and has green, downy stems, and very large, oval, green leaves. Flowers small, orange-yellow. It is a tender species without rhizomes, and requires to be kept constantly growing. Peru.

Canna nepalensis.—A variety of *C. glauca* (according to M. Chaté), introduced into France by M. Wallich, director of the Botanic Garden at Calcutta. About 6½ ft. high, with oval-lance-shaped glaucous leaves, narrowed at both ends. Outer divisions of flower greenish-yellow; inner ones, sulphur-coloured or light orange.

Canna purpurea spectabilis. — Rhizomes cylindrical. Stalks medium-sized, purple. Leaves deep greenish purple, fair size. Flowers small, scarlet. Rather uncommon; habit fine; height 6 ft. to 6½ ft. This is a very hardy species, and has given rise to all the varieties with red or striped leaves.

Canna Warscewiczii.—A very early-flowering species, easily raised from seed, and if sown early in spring, may be used for decorative purposes the same year. Stems numerous, reaching a height of somewhat over 3 ft. Leaves oval-elliptical, narrowed at both ends, and deeply tinged with dark purple. Flowers with brilliant scarlet inner divisions; outer ones purplish. S. America. There are many handsome varieties of this species in cultivation.

Canna Alfred Dumesnil.—Rhizomes clouded, elongated. Stalks medium thickness, deep greenish-purple. Leaves medium-sized, acuminate and erect, deep green striped with violet-purple. Flowers large, well-formed, scarlet. Free-flowering; height 3½ ft. to 4 ft.

Canna Annei.—A vigorous kind, with numerous stiff stems, of a sea-green colour, 6½ ft. high, and large green, glaucescent, oval-acute leaves, 2 ft. long by 10 ins. wide. Flowers large, well-formed, salmon-colour, continuing to appear from July until the first frosts.

Canna Annei-rosea.—Flowers late and scantily. Stems numerous, dark green with a reddish base, attaining a height of nearly 10 ft. Flowers small, carmine rose-colour. Leaves dark green, very narrow and pointed, erect, about 2 ft. 4 ins. long. Rootstock long, conical-cylindrical, reddish.

Canna Annei-bicolor.—A kind with rather thick green stems nearly 6 ft. high, with a violet-coloured base. Leaves light green, oval-acute. Flowers few, of medium size, orange-coloured; the two upper petals reddish. Rootstock of a violet colour.

Canna Annei-floribunda.—Resembles the type in habit and foliage, but is not so tall, is more free-flowering, and has yellow leaves.

Canna Annei-fulgida.—Stems small, of a dark-red colour, from 3 ft. to 5 ft. high. Leaves deep purple, erect, 20 ins. long and 6 ins. wide. Flowers large, well-opened, orange-red. Very tender.

Canna Annei-marginata.—Stems of medium size, dark-red colour, and about 3½ ft. high. Leaves of a deep green with a dark-red margin. Flowers rather large, orange-red. (Considered an inferior variety.)

Canna Annei-discolor.—Stems five or six in number, vigorous and erect, of a uniform red hue, from 3 ft. to 5 ft. high. Leaves lance-shaped, erect, of a light-red colour, 2½ ft. long and 10 ins. wide. Flowers late and few, small, rose-coloured, tinged with yellow. Rootstock conical, very small and short, covered with violet scales.

Canna Annei-rubra.— Rhizomes of good thickness. Stalks medium thickness, purple. Leaves acuminate, deep green striped with purple, good size ; habit fine. Flowers bright orange, well-formed. One of the most free-flowering Cannas. Height 6½ ft.

Canna atronigricans.—Stems of medium thickness, seldom exceeding 3¼ ft. in height. Leaves of a purplish shade passing into dark-red, of a deeper hue than those of *C. nigricans.* Flowers few, of a golden-brown colour. Rootstocks small and few, with reddish scales. A very handsome but tender kind.

Canna aurantiaca-splendida.—Stems green, downy, rather thick, 6½ ft. to over 8 ft. high. Leaves oval, erect, 2 ft 4 ins. to 2½ ft. long and 1 ft. wide, with very prominent membranes. Flowers in September, orange, well-formed, and of a good size. Rootstocks small. A very vigorous-growing variety.

Canna aurantiaca-zebrina.—Stems brown, downy, 3½ ft. to nearly 5 ft. high. Leaves of a light green, striped with fine violet bands, and 2 ft. in length by 10 ins. wide. Flowers very few, light red. Rootstocks few and short.

Canna Amelia.—Stems of a sea-green colour, nearly 5 ft. high. Leaves oval-acuminate, glaucous, and gracefully recurved. Flowers large, well-opened, of a golden yellow, spotted with orange-purple on all the petals. Rootstocks conical and cylindrical. Blooms abundantly and continuously from July till the first frosts.

Canna atropurpurea.—Stems very small and downy, of a reddish hue, and not exceeding 3½ ft. in height. Leaves small, recurved, of an almost black colour. Flowers rather large, of a reddish golden-brown. Rootstocks small, cylindrical, closely crowded around the plant. This variety seeds well and freely.

Canna Abbé Rosier.—Stems green, nearly 6 ft. high. Leaves erect, of a glaucous sea-green shade. Flowers of medium size, light brown, tinged with vermilion, not well-formed, and opening badly. Rootstocks conical and cylindrical.

Canna Bonetti.—Stems vigorous, of a deep-red colour, glaucous towards the top, and about 8½ ft. high. Leaves deep green, with deep-red veins and edges, 2 ft. 4 ins. long, and over 8 ins. wide. Flowers rather large, yellowish-brown, shaded with dark purple, well-formed and well-opened. Rootstocks thick, elongated. Seeds freely and well.

Canna Bonetti-major.—Very like the preceding, but with much taller stems and larger leaves and flowers.

Canna Bonetti-semperflorens.—Stems brown, 6½ ft. to over 8 ft. high. Leaves lance-shaped, deeply tinged and margined with a dark-red shade. Flowers rather large, of a wine-red colour. A very free bloomer.

Canna Bourcier.—Stems brown, 6½ ft. to nearly 10 ft. high. Leaves long, lance-shaped, of a deep violet-red colour. Flowers of a brick-red shaded with salmon-colour. Rootstocks conical, of medium size, violet-coloured.

Canna Bihorelli.—Stems purple, 3¼ ft. to nearly 5 ft. high. Leaves dark red, with a narrow purple margin. Flowers numerous, light red, in handsome panicles. Rootstocks conical and rather numerous.

Canna compacta-grandiflora.—Stems brown, 3¼ ft. to nearly 5 ft. high. Leaves dark red, not well set. Flowers very large, well-formed, of a salmon-shaded-red, in handsome panicles. Rootstocks conical, rather large, covered with reddish scales.

Canna Chatei-discolor.—Stems purple, very robust, 3¼ ft. to nearly 5 ft. high. Leaves thick and firm, of a deep green, finely rayed and margined with purple. Flowers blood-red, well-formed. Rootstocks conical, of medium size. A very shy bloomer.

Canna Chatei-grandis. — Stems brownish-purple, 6½ ft. high, when not divided year after year. Leaves at first erect, afterwards deflected throughout their entire length and spreading very much, 2 ft. 4 ins. long and 1 ft. wide. Flowers late, crimson, in a handsome panicle. Rootstocks brown, conical, very short. Ripens its seeds well.

Canna Daniel Hooibrenk.—Rhizomes large. Stalks strong, green. Leaves large, glaucous-green, acuminate. Flowers large, bright orange. Free-flowering; fine habit. Height over 6 ft.

Canna De Candolle.—Stems green, about 4 ft. high. Leaves green, badly set. Flowers very numerous, vermilion shaded with carmine. Rootstocks conical.

Canna discolor-floribunda.—Stems dark red, 3½ ft. to 4½ ft. high. Leaves oblong, 18 ins. long and 10 ins. wide, deep green striped with purple. Flowers small, orange-red. Rootstocks conical, of medium size. Ripens its seeds well.

Canna Député Hénon.—Stems green, not exceeding 4 ft. 10 ins. in height. Leaves of a light glaucous-green, oval, acute, erect. Flowers large, well-formed, of a pure canary-colour, with a brownish-yellow base, in numerous spikes, which rise gracefully above the foliage to the height of about 20 ins. Rootstocks cylindrical, elongated, standing at some distance from the centre of the tuft. Ripens its seeds well.

Canna Edward Morren.—Rhizomes conical. Stalks green, numerous. Leaves clear green, longish. Flowers large, well-formed, yellow, very much spotted with bright nasturtium-colour. Very floriferous. This is the finest Canna with spotted flowers. Height 5 ft. to 5½ ft.

Canna elata-macrophylla.—Stems reddish below, green and downy above, more than 8 ft. high. Leaves light green, very large, oval, slightly waved. Flowers salmon-coloured, small. Rootstocks conical, of medium size. Seeds freely.

Canna expansa.—Stems thicker than in any other

variety, green, downy, 6 ft. to over 8 ft. high. Leaves of enormous size (over 4 ft. long and from 22 ins. to 2 ft. wide), oval, obtuse, spreading horizontally to a great distance from the stems. Flowers small, vermilion shaded with salmon, in panicles. Roots fibrous. Should be planted in a well-sheltered position to save the huge leaves from being torn by the winds. Never seeds.

Canna expansa-rubra.—Stems numerous, vigorous, very thick, dark-red, 4 ft. to nearly 6 ft. high. Leaves of a dark-red colour, and resembling those of the preceding in size, shape, and arrangement. Flowers large, with rounded divisions, of a dazzling purple colour. Rootstocks very thick, cylindrical, and much swollen in the upper part. Like the preceding, requires a sheltered position.

Canna elongatissima-rustica.—Stems deep-green, very thick, 6½ ft. to over 8 ft. high. Leaves oval, erect, very large, of a deep shining green, with a narrow dark-red margin. Flowers very small, orange-rose-coloured. Rootstocks very small, conical, with fibrous roots. A vigorous-growing variety.

Canna excelsa-zebrina. — Stems dark violet-red, downy, rather thick, 6½ ft. to over 8 ft. high. Leaves very large, oval, erect, deep-green, passing into dark-red, rayed with violet-purple. Flowers small, orange. Rootstocks small, conical. This is the finest striped variety.

Canna guayaquilla.—Rhizomes large, round. Stalks very thick, reddish. Leaves very wide and large, bordered with purple. Flowers small, yellowish - orange. Rather uncommon; fine habit; a remarkable kind. Height over 6 ft.

Canna gigantea-major.—Stems thick and vigorous, of a light-green, slightly reddish below, 6½ ft. to 8½ ft. high. Leaves very large, of a light glistening green. Flowers medium size, pale-red. Rootstocks whitish, very thick, cylindrical, and swollen. A very hardy variety.

Canna gigantea-rubra.—Stèms dark red, 4 ft. to nearly 6 ft. high. Leaves broad, obtuse, green, shaded with dark-red. Flowers blood-red, in panicles. Rootstocks conical, very closely crowded together. Resembles *C. Chatei-grandis*, but is an inferior variety.

Canna grandis.—Stems green, downy, very thick, from nearly 10 ft. to nearly 12 ft. high. Leaves oval, erect, deep green, very large. Flowers poppy-red, small, but in large panicles. Rootstocks small, with fibrous roots.

Canna grandiflora-floribunda.—Stems small, from 20 ins. to 3½ ft. high, of a light glistening green. Leaves small, light-green and glistening. Flowers numerous, very large, with rounded divisions, orange-rose colour, in handsome panicles. Rootstocks yellowish, numerous, cylindrical, of medium size. A very hardy variety, of great effect from its brilliant and luxuriant inflorescence.

Canna Heliconiæfolia.—Stems deep green and downy, 6½ ft. to over 8 ft. high. Leaves green, oval, erect, very firm. Flowers small, orange. Roots fibrous. Does not seed before the second year.

Canna Hostei.—Stems chestnut-coloured, about 8 ft. high. Leaves dark-red, lance-shaped. Flowers large, red. Rootstocks dark-red, conical.

Canna involventiafolia.—Stems green, 8 ft. to nearly 10 ft. high. Leaves very large, reflexed, of a light green. Flowers few and small. Rootstocks small, with fibrous rootlets.

Canna Imperator. — Stems vigorous, very thick, green and downy, reddish below, 6½ ft. to over 8 ft. at the close of the season. Leaves half-opened, lance-shaped, light green, with prominent lateral nerve, 2 ft. 4 ins. long by 14 ins. wide. Flowers late, very effective, of a dazzling blood-red. Rootstocks conical, very short.

Canna iridiflora-hybrida. — Stems green, downy, somewhat reddish below, 6½ ft. to over 8 ft. high. Leaves green, very large. Flowers well-formed, very large, blood-red. Rootstocks small, with fibrous roots. Flowers to most advantage in a house, where it is really magnificent.

Canna iridiflora-rubra. — Stems brown, 3½ ft. to nearly 5 ft. high. Leaves deep green shaded with dark red, and with a narrow dark purplish-red margin. Flowers large, purplish-red. This variety is not so tender as the preceding one.

Canna insignis. — Stems violet, downy, 3½ ft. to nearly 5 ft. high. Leaves oval, extending horizontally, of a tender green, rayed and margined with purplish-red. Flowers few and small, of an orange-red. This variety is valuable for its fine foliage.

Canna Joseph-Auzende.—Stems green, 4½ ft. high. Leaves green and flaccid. Flowers deep carmine, of medium size. Rootstocks yellowish, small and conical. Not a very good variety.

Canna Jean Bart. — Rhizomes conical; stalks medium thickness, deep green, elongated. Flowers very large, deep purple; height 5½ ft. to 6 ft.

Canna Jean Vandael. — Rhizomes cylindrical,

longish; stalks medium thickness, numerous, clear green. Leaves clear green, erect, elongated. Flowers large, well-formed, garnet-red; height 4 ft. to 4½ ft.

Canna Jussieu.—Stems green, from nearly 4 ft. to 4½ ft. high. Leaves small, glaucous-green. Flowers citron, approaching a chamois-colour. Rootstocks conical cylindrical. (An inferior variety.)

Canna Krelagei discolor.—Stems very thick, from nearly 5 ft. to nearly 6 ft. high. Leaves broad and thick, dark-red, rayed with purplish-red. Flowers large, carmine-red, sometimes rayed with white, in handsome panicles. Rootstocks grey, conical.

Canna Liervalii.—Stems dark-red, 6½ ft. high. Leaves of the same colour, rayed with purple. Flowers orange-red. Rootstocks dark red, conical.

Canna Lavallei.—Stems slender, purplish-brown. Flowers very large, well-formed, vermilion-orange, becoming yellower as they open. Rootstocks cylindrical.

Canna Lemoinei.—Stems green, 6½ ft. to over 8 ft. high. Leaves very leathery, of medium size, and deep green colour. Flowers bright orange.

Canna limbata-major.—Stems green and downy, 5 ft. to 6½ ft. high. Leaves large, lance-shaped, narrow, spreading, deep green, 2½ ft. long, and over 8 ins. wide. Flowers large, orange-red. Rootstocks cylindrical, narrowed towards the top, closely crowded together. Ripens its seeds freely.

Canna maxima.—Stems green and downy, 5 ft. to 6½ ft. high. Leaves very large, lance-shaped, acute, light green, becoming darker in the course of growth, reflexed, from 2½ ft. to 32 ins. long, and 10 ins. to 1 ft. wide, on stalks 6 or 7 ins. long. Flowers small, orange-yellow. Roots fibrous, without rootstocks. Does not flower before the second year, and is valuable only for its foliage.

Canna Maréchal-Vaillant.—Stems robust, from 5 ft. to 6 ft. high. Leaves oblong, lance-shaped, erect, 28 ins. to 32 ins. long, deep green, striped with purplish-red. Flowers large, elegant, of a pure orange, in handsome spikes. Rootstocks rather thick, conical and cylindrical, standing at some distance from the tuft.

Canna musæfolia-peruviana.—Stems green and downy, 5 ft. to 6½ ft. high. Leaves very large, wide, light green. Flowers small, orange. Rootstocks very small.

Canna musæfolia minima.—Leaves of a whitish green, badly set. Flowers small, orange-brown. No rootstocks.

Canna musæfolia-hybrida.—Resembles *C. musæfolia*, but the stems and leaves are thicker and of a deeper green.

Canna musæfolia-rubra.—Stems dark red, 6½ ft. high. Leaves dark purple-red, oval, very large. Flowers small, salmon-red. Rootstocks very tender, with fibrous roots. Neglected as too tender and not sufficiently distinct from several other varieties, such as *C. Chatei-grandis*, etc.

Canna musæfolia-perfecta.—Stems from 5 ft. to 6½ ft. high. Leaves broad, very firm, of a handsome whitish green. Flowers small, yellow. Roots fibrous, without rootstocks.

Canna metallica.—Resembles *C. nigricans*, and has never justified its name.

Canna macrophylla-zebrina.—Stems violet and downy, 4 ft. to over 5 ft. high. Leaves dark-red, rayed and striped with purple. Flowers red, with a perennially withered aspect. Rootstocks very small, conical, closely set round the tuft. This variety is very subject to diseases.

Canna metallicoides.—Stems violet, 5 ft. to 6½ ft. high. Leaves of medium size, dark-red striped with purple. Flowers medium size, light-red. Rootstocks small, with fibrous roots.

Canna nervosa.—Stems reddish, 3¼ ft. to 5 ft. high. Leaves deep green, rayed and margined with purple. Flowers small, blood-red. Rootstocks conical.

Canna nigricans.—Stems purplish-red, 4¼ ft. to over 8 ft. high. Leaves lance-shaped, acuminate, erect, of a coppery red, which exhibits a metallic gleam under sunshine. The old leaves lose their reddish tint, and assume a duller hue. They are 2½ ft. in length by 10 ins. or 12 ins. wide. Flowers few and late, of a sad, tawny-yellow colour. Rootstocks brownish, conical. Seldom seeds. One of the finest kinds.

Canna nana-superba.—Stems green, growing very closely together, and from 20 ins. to 32 ins. high. Leaves small, of a handsome green. Flowers large, badly formed, of a brick-red, becoming darker as they open. Rootstocks grey, small, conical. (A very inferior variety.)

Canna nepalensis-grandiflora.—Has the same habit and foliage as the type (*C. nepalensis*), but is a dwarfer variety with better-shaped flowers of a sulphur-yellow, sometimes dotted with red.

Canna Oriflamme.—Rhizomes conical, stalks deep

green. Leaves good size, elongated, acuminate, deep green lightly striped. Flowers very large, deep orange, in panicles which stand up considerably above the foliage. Height 5 ft. to 5½ ft.

Canna Pie IX.—Stems small, green, slightly reddish below, 3½ ft. to 4 ft. high. Leaves small, of a pale glaucous-green, erect, lance-shaped, acuminate. Flowers large, light yellow with a deep orange claw, very numerous, in closely-crowded panicles. Rootstocks conical and cylindrical. Excellent for edgings.

Canna Parmentier.—Stems small, green, 4 ft. to 5 ft. high. Leaves glaucous-green, small. Flowers brownish-yellow. Rootstocks grey, conical and cylindrical. (An inferior variety.)

Canna purpurea-hybrida.—Stems dark-red, from 4½ ft. to over 5 ft. high. Leaves dark-red, rayed with reddish-purple. Flowers large, brownish-yellow. Rootstocks small, conical, crowded. A very handsome variety, but tender.

Canna Porteana.—Stems small, dark-red, 3½ ft. to 4 ft. high. Leaves red, with a metallic lustre. Flowers medium-sized, light orange. Rootstocks grey, small, forming a close, compact tuft. A fine but tender variety.

Canna picturata-fastuosa. — Stems numerous, green, 5 ft. to 6½ ft. high. Leaves green, glaucescent, narrow, much pointed, over 2 ft. long by 6 ins. wide. Flowers large, well-opened, light yellow speckled with red. Blooms luxuriantly and continuously from the end of July to the first frosts. Rootstocks white, with grey scales, cylindrical, much elongated.

Canna picturata-nana.—Stems small, green, from

20 ins. to 2 ft. high. Leaves very small, of a light green. Flowers large, yellow, spotted with red. Rootstocks whitish, small, conical. A charming variety for edgings.

Canna Plantieri.—Stems very thick, reddish below, deep green above, 3½ ft. to over 8 ft. high. Leaves green, glaucous, lance-shaped, acute. Flowers large, bright yellow, changing to orange. Rootstocks greyish, rather thick, conical and cylindrical. A late and shy bloomer.

Canna Prémices-de-Nice.—Stems and leaves like those of *C. Annei*, 3½ ft. to 5 ft. high. Flowers very large, bright yellow, sometimes dotted with salmon-colour. Rootstocks conical and cylindrical, rather shorter than those of *C. Annei.* A very free-flowering and hardy variety.

Canna rubra-superbissima.—Stems dark purple-red, very thick, from nearly 6 ft. to 6½ ft. high. Leaves broad, round, purplish-red, with a metallic lustre. Flowers of medium size, light orange-red. Rootstocks brownish, very thick, conical, crowded together. One of the finest metallic-red-leaved Cannas.

Canna rubricaulis.—Stems dark-red, from nearly 6 ft. to 7½ ft. high. Leaves dark-red, rayed and margined with purple-red. Rootstocks greyish, small, conical. (Inferior to the preceding variety.)

Canna rubra-nerva.—Stems dark-red, from 3½ ft. to 5 ft. high. Leaves long and narrow, lance-shaped, reflexed, dark-red rayed with purple. Flowers large, of a cinnabar red. Rootstocks dark-red, very long, conical, and cylindrical. Resembles a weeping willow.

Canna rubra-perfecta.—Stems dark-red, from 5 ft.

to 6½ ft. high. Leaves dark-red, rayed with purple. Flowers of medium size, orange-red. Rootstocks dark-red, conical. Seeds freely.

Canna rotundifolia-vera.—Stems very thick, green, downy below, from 3½ ft. to 5 ft. high. Leaves round, reflexed, of a deep glistening green. Flowers medium-sized, opening very badly, of a carmine-red. Rootstocks medium-sized, conical. (A tender variety.)

Canna rotundifolia-rubra-major.—Stems dark-red, numerous, very thick, vigorous, from 3½ ft. to 5½ ft. high. Leaves round, obtuse, spreading almost horizontally, pale-red, with a dark purplish-red midrib and border. Rootstocks dark-red, very thick, cylindrical, much swollen above.

Canna rotundifolia-metallica.—Similar to the preceding, but with coppery-red leaves which have a metallic lustre.

Canna Rendatleri. — Stems light purplish-red, of medium thickness, vigorous, rather distant from each other, from nearly 6 ft to over 8 ft. high. Leaves much pointed, deep-green tinged with dark-red, badly set. Flowers numerous, very large, well-opened, salmon-red. Rootstocks greyish, conical, cylindrical. Ripens its seeds badly.

Canna Rodezii.—Stems small, numerous, from 4 ft. to 6½ ft. high. Leaves green, glaucous, lanceolate, acute, very narrow. Flowers large, marigold-orange. Rootstocks whitish, cylindrical, very slender and long. A variety desirable for its flowers, but not ornamental in foliage.

Canna Thibauti.—Stems purplish-brown, from 5 ft.

to 6½ ft. high. Leaves broad, thick, dark-red, rayed with purple. Flowers very large, well-opened, poppy-red, in crowded panicles. Rootstocks white, small, crowded together.

Canna striata.—Stems green, slender, 3¼ ft. to 5 ft. high. Leaves small, very narrow, erect, glaucous-green. Flowers yellow, speckled and rayed with red. Rootstocks small, cylindrical, crowded at the base of the plant. Inferior to *C. picturata-nana* of the same colour.

Canna Warscewiczioides-Chatei. — Stems dark-red, thick, from 6½ ft. to over 7 ft. high. Leaves very large, dark-red. Flowers small, blood-red, in very large panicles. Rootstocks brown, rather thick, cylindrical.

Canna Warscewiczioides-nobilis. — Stems deep-green, tinged with dark-red, from 5 ft. to 6½ ft. high. Leaves deep-green, rayed and margined with dark-red. Rootstocks medium-sized, conical, crowded around the base of the plant.

Canna Van-Houttei.—Stems dark-red, vigorous, from 5 ft. to 6½ ft. high. Leaves lance-shaped, pointed, green, rayed and margined with dark purplish-red, 2 ft. to 2½ ft. long. Flowers large, well-formed, poppy-red. Rootstocks grey, long, cylindrical. A very handsome and hardy variety.

Canna zebrina.—Stems green, tinged with dark-red, small, from 32 ins. to 3¼ ft. high. Leaves deep-green, rayed and striped with dark-red. Flowers small, orange tinged with salmon. Rootstocks whitish, conical. A tardy grower.

Canna zebrina-major.—Stems green tinged with dark-red, downy, of medium size, 5 ft. to 6½ ft. high.

Leaves oval, erect, deep-green, rayed and striped with dark purplish-red. Flowers very small, red. Rootstocks brown, rather thick, conical, swollen.

Canna zebrina-elegantissima.—Stems dark-red, vigorous, 4 ft. to 5 ft. high. Leaves very large, lance-shaped, deep-green, striped with dark purplish-red. Flowers of medium size, yellowish-brown tinged with scarlet. Rootstocks greyish, conical, swollen.

Canna zebrina-géant.—Stems deep-green, tinged with violet-red, very thick, downy. Leaves very large, thick, deep-green, rayed with dark-red. Flowers large, light-red. Rootstocks whitish, conical, closely crowded together. Difficult to flower, but very ornamental in foliage.

Canna zebrina-violacea.—Stems violet, downy, slender, about 4 ft. high. Leaves violet, rayed with purple. Flowers of medium size, bright-red. Roots fibrous. A very tender variety, with badly-developed leaves.

Canna zebrina-nana.—Stems green tinged with dark-red, 16 ins. to 20 ins. high. Leaves small, light-green, rayed and margined with purple. Flowers of medium size, light-red. Rootstocks small, crowded on one another. Excellent for large edgings.

Canna Ferrandii.—Stems dark purplish-red, 4 ft. to 5 ft. high. Leaves medium-sized, dark-red, margined with purple. Flowers large, blood-red, very numerous, in handsome panicles.

Canna Auguste Ferrier.—Stems green, very thick, downy, nearly 10 ft. high. Leaves very large, oval, erect, pointed, deep-green, with narrow stripes and margin of dark purplish-red. Flowers of medium size, orange-red.

Rootstocks small, but very hardy. A shy bloomer, but of remarkably fine habit and foliage.

Canna Barilletti.—Stems red, vigorous, nearly 10 ft. high. Leaves very large, deep-red. Flowers small, orange-red. Rootstocks very small, conical, with fibrous roots.

The number of Cannas enumerated is so large that it is desirable to make a selection from them. A still narrower selection is indicated by asterisks.

A SELECTION OF CANNAS.

C. Annei
,, -discolor
atronigricans
atropurpurea
*Auguste Ferrier
aurantiaca-splendida
aurantiaca-zebrina
*Amelia
Barilletti
Bihorelli
Bonetti
*Chatei-discolor
,, -grandis
Député Hénon
discolor
*excelsa-zebrina
expansa
,, -rubra

C. *elegantissima rustica
Ferrandii
gigantea
grandiflora-floribunda
iridiflora
,, -hybrida
,, -rubra
insignis
*Imperator
*Krelagei discolor
Lavallei
Liliiflora
limbata
macrophylla
musæfolia
maxima
Maréchal-Vaillant
*nigricans

C. peruviana purpurea
,, robusta
*purpurea spectabilis
*Porteana
*picturata-fastuosa
,, -nana
Pie IX.
Prémices-de-Nice
Rendatleri
rotundifolia-metallica
,, -rubra
Thibauti
Van-Houttei
zebrina-nana
,, -elegantissima

***Cannabis sativus** (*Hemp-plant*).—A well-known annual, native of India and Persia, and largely cultivated in Europe for the sake of its fibre. In ordinary situations it grows from 4 ft. to 10 ft. high, but in Italy, under very favourable circumstances, it sometimes grows as high as 20 ft. In plants growing singly the stem is frequently much branched, but when grown in masses it is generally straight and unbranched. The leaves are long-stalked, and composed of from five to seven long, lance-shaped, sharp-pointed leaflets, radiating from the top of the stalk, and with the margins cut into sharp saw-like teeth. This well-known plant is useful where the tenderer subtropical plants cannot be enjoyed. Single well-grown plants of it look very imposing and distinct, and are good for the backs of borders or mixed groups. For these purposes, it should be sown early in April in the open ground. To get large plants it would no doubt be worth while raising it in frames. It loves a warm, sandy loam.

***Carduuse riophorus** (*Woolly-headed Thistle*).—A remarkably conspicuous native plant, with a much-branched, furrowed, hairy stem 3 ft. to 5 ft. high, and very deeply cut and undulated spiny leaves, the lower ones often 2 ft. long. The flower-heads are very large, of a purplish-red colour, and surrounded on the under side with a dense white cottony web. There are few plants more handsome or novel in appearance than an established one of this. It is suitable for borders, or groups of hardy fine-foliaged plants, and grows well in any ordinary garden-soil. Seed.

***Carex paniculata.**—A very large sedge, growing somewhat like a dwarf tree-fern, with strong and thick stems, and with luxuriant masses of drooping leaves. The roots form dense elevated tufts, frequently elevated from one to three feet above the surface of the ground; and when the plant is in flower, it generally exhibits a large and spreading panicle. The leaves are rough and broad, and the flower-spike from 3 ins. to 6 ins. long. A few tufts of this are very effective on the margins of water near groups of picturesque plants. The finer specimens are of great age, and must be procured from the bogs where the plant occurs wild.

***Carex pendula.**—A very handsome plant, unlike any of the other British Carices, growing in large round tufts, with numerous flowering-stems and barren shoots, which attain a height of from 3 ft. to 6 ft. The leaves are often 2 ft. or more in length, and are chiefly at the base of the plant. It is most attractive when in flower, from the graceful disposition of its pendent spikes, which are usually about half-a-dozen in number, and each from 4 ins. to 7 ins. in length. Very suitable for the margin of water or for boggy or moist spots.

***Carlina acaulis.**—A hardy perennial, rather interesting from its foliage, which has some resemblance to the leaves of a miniature Acanthus, and is disposed in a broad, handsome, regular rosette very close to the ground. Its single yellowish flower, 3 ins. or more across, is borne on a very short, erect stalk in the centre of the rosette. Although too dwarf for association with plants of more imposing stature, it is well worthy of a place on a bank or slope, or on the margins of low beds or groups, where its pleasing aspect and very distinct habit will be

Carlina acaulis.

CARYOTA SOBOLIFERA

Tender Palm; for summer use in the southern counties only.

seen to best advantage. It thrives best in dry, stony, calcareous soil, and is easily multiplied by sowing. In the mountainous districts of France the flowers are gathered by the inhabitants, and used as a substitute for artichokes. Central Europe.

Carludovica palmata.—A very ornamental, palm-like plant, from 4 ft. to 7 ft. high, with rich dark-green leaves from 2 ft. to 3 ft. broad, and divided into four lobes, each of which is again divided at the apex into narrow segments. The leaf-stalks are round, smooth, and without spines, and are of the same colour as the leaves. This interesting plant will stand the open air in summer, from early June till October, but requires warm house treatment in winter, with plenty of water at all times. Seed. Peru and New Granada.

Caryota sobolifera.—An elegant Palm, with a slender stem and shining light-green bipinnate leaves. The leaf-stalks, when young, are clothed with a short, black, scaly tomentum. which falls off as the plant grows older. It is often confounded with *C. urens*, but may be easily distinguished from it by the suckers which it produces very freely from its base. Similar treatment and uses to those given for *C. urens*, with which it is of much the same value for the open garden. Malacca.

Caryota urens.—An East Indian Palm, with a stout stem, and an elegant crown of spreading bipinnate leaves, from 3 ft. to 12 ft., or more, in length, of a dark-green colour, the leaflets being 6 ins. to 9 ins. long by 2 ins. to 4 ins. wide. When young, it should be potted in equal parts of loam and vegetable mould, with a little sand; the pot to be well drained and water given liberally during the growing season. It is generally seen in a small state in this country; and though it stands the open air in summer, from June till the end of September, pretty well, it can never be of much importance for our open-air gardening.

***Cassia marilandica.**—A hardy, graceful perennial, 3½ ft. to 5 ft. high, with pinnate leaves, resembling those of the Acacia, and slender stems, bearing yellow flowers, in numerous small clusters in autumn. It is somewhat late in growth, but once commenced, grows with great rapidity. It thrives best in a position with a south aspect, and may be multiplied either by division in spring, or by sowing from April to June. It should always be planted in a warm, deep, sandy loam, and is very suitable for borders or association in groups with the finer hardy subjects, its graceful leaves qualifying it for a place in a group of hardy foliage-plants. In naturally warm, deep, and well-drained soils it will prove a noble subject for the back parts of borders. N. America.

***Centaurea babylonica.**—Among the Centaureas there are a few subjects which might be used among hardy fine-leaved plants, but by far the most distinct and remarkable is the very silvery-leaved *C. babylonica.* This is quite hardy, and when planted in good ground, sends up strong shoots, clad with yellow flowers, to a height of 10 ft. or 12 ft. The bloom, which continues from July to September, is not by any means so attractive as the leaves; but the plant is at all times picturesque. In groups. or, still better, isolated, on rough or undulating parts of pleasure-grounds, it has a very fine effect. A free sandy loam suits it best. Seed. Levant.

CENTAUREA BABYLONICA

Coarse herbaceous Section; for isolation by wood-walks, etc.

Centaurea Clementei.—A plant of robust growth (resembling *C. ragusina*, but much larger in every part), with broad crowns of leaves, which are deeply serrated and cut into lobes. To the elegance of the foliage must be added its beauty of colouring, the leaf in a young state being covered with down as white as snow, and when fully matured and developed still retaining a silvery appearance. This plant, which I first saw in M. Boissier's garden, near Lausanne, I have no experience of as a hardy plant in this country; but whatever its value in this respect may be, there can be no doubt that for the summer garden it will prove as effective as either *C. gymnocarpa* or *C. ragusina*, both exceptionally fine and useful plants. The same treatment will suit it.

Centaurea dealbata is a dwarf hardy species from the Caucasus, with elegant foliage, white underneath, and rosy flowers: suitable among dwarf hardy subjects.

Centaurea gymnocarpa.—A half-shrubby plant from the South of Europe, nearly 2 ft. high, with hard, branching, bushy stems, and elegantly cut, arching leaves, which are covered with a short, whitish-satiny down. A variety (*C. plumosa*) has the leaves much more divided, and not so white. This plant is somewhat hardier than *C. ragusina*, but both require greenhouse treatment in winter. Same soil, positions, and treatment as for *C. ragusina*. Useful as this is as an edging or bedding plant, it is when grown as fine single specimens that its beauty is most seen.

Centaurea ragusina.—This fine and distinct plant, which has lately become one of the most popular of our flower-garden ornaments, cannot be passed by in a book on fine-leaved plants. It is so abundantly used as a bedding and ribbon-plant, etc., that nothing need be said of it in these respects, but it will be seen to great advantage in single, well-grown tufts or small groups, as its silvery leaves would contrast finely with many of the dark green and glossy things recommended for this purpose. Readily increased by seeds, or by cuttings. Should be planted out in May. S. Europe.

Chamædorea.—A family of Mexican palms, with smooth, fine green stems, resembling those of the bamboos, seldom more than 15 ft. or 20 ft. high and 1 in. or 2 ins. thick, surmounted by tufts of eight or nine pinnate or almost entire leaves, nearly 8 ft. in length. Among the most ornamental species are *C. elatior*, *C. elegans*, and *C. Ernesti-Augusti.* These elegant palms may, with advantage, be placed in the open air in early summer, in sunny but sheltered nooks, and taken in at the end of September. Their small, elegant heads particularly fit them for placing here and there among groups of medium-sized, fine-leaved plants, or among mixed masses of dahlias, cannas, etc.

***Chamæpeuce diacantha.**—A spiny-leaved biennial of remarkable habit, growing in tufts of six or eight stems, from 2 ft. to nearly 3 ft. high, which, in the following season, are terminated by dense, spike-like clusters of purplish blooms. It requires light, well-drained soil and a warm position, and should seldom be watered. As the stems are not produced until the second year, the radical rosettes of the first year may be advantageously used in forming edgings, or on the margins of groups, for which

CHAMÆDOREA.

Slender Palm Type; for placing amidst groups of dwarfer subjects during the summer months.

their light-green, silver-veined leaves are very suitable, or they may serve to fill a vacant space in the mixed border. Multiplied by sowing in spring or autumn. The plants require the shelter of a house in winter, and are only effective for edgings in the young or rosette stage.

***Chamærops excelsa.**—A hardy species, with an erect stem, 20 ft. or 30 ft. high in its native country, and dark-green, erect, fan-shaped leaves, deeply cut into narrow segments. The leaf-stalks are from 3 ft. to 6 ft. long, and are enclosed at the base in a dense mass of rough fibres, and armed at the edges with small, tooth-like spines. This plant remains out during the winter in the neighbourhood of Paris, in sheltered positions, the stem being protected in severe frosts with a covering of straw, etc., and it is worth a trial in the south with us.

***Chamærops Fortunei** (*The Chusan Palm*).—This species is often confounded with *C. excelsa*, from which, however, it differs in being of a stouter habit, having a more profuse matted network of fibres around the bases of the leaves and crown, the segments of the leaves much broader, and the leaf-stalks shorter and stouter, from 1 ft. to 2 ft. long, and quite unarmed. It grows 12 ft. or more in height, and has a handsome, spreading head of fan-like leaves, which are slit into segments about half-way down.

It may not be generally known that this palm is perfectly hardy in this country. A plant of it in Her Majesty's gardens at Osborne has stood out for many winters and attained a considerable height. It is also placed out at Kew, though protected in winter. On the water-side of the high mound in the Royal Botanic Gardens, Regent's Park, it is in even better health than at Kew, though it has not had any protection for years, and stood the fearfully hard frost of 1860. If small plants of this are procured, it is better to grow them on freely for a year or two in the greenhouse, and then turn them out in April, spreading the roots a little and giving them a deep loamy soil. Plant in a sheltered place, so that the leaves may not be injured by winds when they grow up and get large. A gentle hollow, or among shrubs on the sides of some sheltered glade, will prove the best place for it. The establishment of a palm among our somewhat monotonous shrubbery and garden vegetation is surely worthy of a little trouble, and the precautions indicated will prove quite sufficient.

Chamærops Palmetto (*Palmetto Palm*).—This is a rather slow-growing species, but valuable on account of its hardiness. It grows to a height of about 15 ft., and has glaucous or sea-green, fan-shaped leaves, divided into long narrow segments. The stem is smooth or without prickles. It is a very fine object when planted out; and, grown in tubs in a cool house or conservatory, stands the open air in summer well, and should be put out at the end of May.

***Chenopodium Atriplicis.**—A vigorous-growing Chinese annual, with an angular, erect, slightly branched, reddish stem, over 3 ft. in height, the young shoots and leaves covered with a fine rosy violet powder. The leaves are very numerous, nearly spoon-shaped, and long-stalked. This plant is very ornamental in foliage, and well adapted for planting on grass-plats or grouping with

CHAMÆROPS EXCELSA.

Hardy Palm: best in sheltered positions.

other plants in pleasure-grounds. May be treated as a half-hardy annual.

Cineraria maritima.—A very handsome bushy perennial, with finely-cut leaves, covered on the under side with a silvery down. It bears numerous heads of bright yellow flowers in summer. When the effect of its foliage only is desired, the flowering-stems should be pinched off on their first appearance. The plant then becomes more leafy and more branching. Multiplied easily by seeds. Useful on the margins of shrubberies, or isolated on banks, or on the grass of the pleasure-ground, where it would form an agreeable variety among the *Acanthuses* and various other dark-green subjects recommended for this purpose.

Cineraria Acanthifolia is a fine variety of the preceding, and well worthy of cultivation.

Cladium Mariscus.—This vigorous native fen-plant grows from 2 ft. to 6 ft. high, and, when in flower, is crowned with dense, close, chestnut-coloured panicles, which are sometimes 3 ft. in length. The radical leaves are glaucous, rigid, and often 4 ft. long. Worthy of a place near such subjects as *Carex pendula* or the *Typhas* on the margin of water.

Colea Commersonii.—A shrub from Madagascar, with very large opposite leaves, each consisting of pairs of oval-obtuse leaflets. This has been used in the subtropical garden at Battersea Park, but it is not likely to be of any practical importance in our outdoor gardening. Indeed it must be a very fine plant in this way which would produce so good an effect as young plants of *Ailantus glandulosa.*

***Comptonia asplenifolia.**—This is a quaint-looking little shrub, fern-like in leaf and neat in habit. The leaves are long, and cut into numerous rounded lobes, somewhat like those of the *Ceterach;* and the habit of the shrub is spreading and pleasing. It likes peaty soil, and may be increased by layers, suckers, or seeds. It should be used as an isolated specimen on the grass, or associated with such things as the oak-leaved *Hydrangea* and *Aralia japonica.* A very pretty object in the woods of New Jersey and many other parts of N. America, where it is called the Sweet Fern. It would be peculiarly appropriate for similar positions in this country, when we have it plentiful enough to naturalise.

Cordyline indivisa.—A magnificent New Zealand plant, with a simple stem, from 2 ft. to 5 ft. high, and well distinguished by its excessively thick and leathery leaves, which are from 1 ft. to 5 ft. long, and 4 or 5 inches broad, lance-shaped, of a dark shining green colour, the midrib and veins being of a rich deep orange. The flowers are white, and very densely crowded, in a large drooping panicle. This fine greenhouse plant may be placed out of doors in summer, from the end of May till October, with a very good effect; best, perhaps, as an isolated specimen, the pot being sunk in the grass.

Corypha australis.—A noble Australian palm, over 30 ft. high in its native country, and forming a very effective subject for the subtropical garden in summer, from June till October. The leaves are nearly circular, often more than 5 ft. broad, of a dark green colour, very much plaited, and divided round the edge into narrow segments, and supported by spiny leaf-stalks, from 6½ ft. to nearly 10 ft. long. It requires abundance of water, and should have a warm, sunny, and sheltered position. Few places, however, can afford to have subjects of this character in the open air, except where there are large conservatories, in which it is a relief to get more room in summer.

***Crambe cordifolia.**—This is unquestionably one of the finest of perfectly hardy and large-leaved herbaceous plants. It is as easily grown as the common Seakale—more easily, if anything; and in heavy rich ground makes a splendid head of leaves, surmounted in summer by a dense spray of very small flowers. In planting it, the deeper and richer the soil the finer the result. It will prove a capital thing for any group of fine-leaved hardy plants, and may also be planted wherever a bold though low type of vegetation is desired.

There is another species, **C. juncea*, a dwarf kind, with white flowers and much-branched stems, the ramifications of which are very slender and elegant. This is also effective, but not so valuable as *C. cordifolia.*

***Cucurbita perennis.**—A climbing or trailing plant, well adapted for covering walls, ruins, trellises, steep slopes, etc. It is a very vigorous grower, its stems sometimes attaining a length of nearly 40 ft. in one year; but it will probably never do this in our climate. The leaves are strong, rough, and of a glaucous colour; and the shoots run about freely if the plant be in very rich soil. Where a bold trailing plant for high trellis-work, or rough banks, or shaggy rockwork is desired, it will be found useful; but withal we cannot give it a place in the front rank, and the small select garden without any of the above-mentioned appendages will certainly be better without it. For the botanical garden and curious collections it is indispensable. It is strong and lasting when well established, and may be allowed to trail over rough places, stumps, or similar positions. The flowers have a rather strong odour of violets, and are succeeded by roundish fruit, the size of a small orange, of a deep green colour barred and speckled with white. Requires a deep, light soil, and a warm but airy position. Seedlings and plants in exposed places should be covered in winter with litter or leaves. It is easily multiplied in spring by division of the young tuberous stems, or by sowing in light, substantial, well-drained soil from April to July, or in pots in a lukewarm hotbed in March and April. It has not, so far as I am aware, fruited in our climate.

Cyathea dealbata.—This very handsome fern, known in N. Zealand as the Silver Tree-fern, has a slender, branched, almost black stem, 4 ft. to 8 ft. high, ending in a fine crown of broadly-oblong twice-divided fronds of a dark green colour above and milk-white below; the rachis and midribs when young are covered with brown scales, and afterwards with pale deciduous down. This plant may be placed in the open air, in the southern and milder districts, during the summer months from the end of May till the end of September.

Cycas revoluta.—A graceful and well-known plant, with a very stout stem, sometimes, though rarely, reaching a height of from 6 ft. to 10 ft., from the top of which issues a beautiful crown of dark green pinnate

CYCAS (*very large and old specimen*).

Stove Section: suitable for placing in the open air, in warm and sheltered parts of the country, after a strong growth has been made and matured indoors.

leaves, from 2 ft. to 6 ft. long. It is one of the most valuable of the greenhouse plants that may be placed in the open air in summer from the end of May till October, and is particularly graceful in the centre of a bed of flowering plants, or isolated with the pot or tub plunged to the rim in the turf, always in a warm and sheltered position. Increased by seeds, or by separation of the suckers which are occasionally thrown up.

***Cynara Scolymus** (*French Artichoke*).—This plant, although chiefly grown for culinary purposes, possesses sufficient merit as a foliage-plant to entitle it to a place amongst ornamental subjects. Its long, deeply-divided leaves, white and downy beneath, its height (4 ft. to 5 ft.), its purplish flower-heads, and distinct habit render it very suitable for planting on the irregular and rougher parts of pleasure-grounds, grass-plats, etc., which are often occupied by subjects far less striking.

***Cyperus longus** (*Galingale*). — The stiff, erect, tapering, triangular stem of this plant, which is from 2 ft. to 3 ft. high, is crowned by a handsome loose umbellate panicle of chestnut-coloured flower-spikes, at the base of which there is an involucrum of three or more unequal leaves. These are often 1 or 2 feet long, the lower ones arching gracefully and of a bright shining green, giving the plant a very distinct and pleasing appearance. The rootstock is thick and aromatic, and was formerly much used in medicine as a tonic. A rare native plant, suitable for the bog-bed or the margin of water.

Dahlia imperialis (*Lily-flowered Dahlia*). — The common Dahlia gives us no more idea of this than the little vernal Scilla of Britain does of *Scilla peruviana.*

The Imperial Dahlia has very large and graceful, much divided leaves, and flowers of a pure and beautiful French white, thrown up in a great cone-like mass, and resembling such lilies as *L. tigrinum Fortunei*, which, instead of merely developing a head of flowers, shoot up a great candelabrum laden with them. The flowers of this dahlia do not, like most of the flowers of composite plants, open so wide as to stare at you with the brazen look of a sunflower, but, on the contrary, hang pendulous and half open, with some of the modesty of the white lily. There is little chance of this species producing its flowers in the open air in this country, but it will, notwithstanding, be of service both in the flower-garden and conservatory.

Planted in rich soil, and placed in a warm, sheltered position in the open air at the end of May, it grows well with us in summer, and, in consequence of its large and graceful leaves, is an ornament worthy of being used as a "fine-foliaged" or "subtropical" plant. Just at the time that it begins to gather together its flowering energies the best of our season fails, and the plant must soon fail too, if not immediately taken up and placed in a well-lighted and warm greenhouse. If plunged out in a large pot or tub during the summer, it may be taken up without injury, and will in all probability flower under glass in the autumn, and prove a magnificent ornament. We should have very small hopes of its flowering well if planted out so that its great roots must be mutilated when being taken up, and therefore the safe way will be to pot it in a very large pot, and plunge that in the ground. The roots would probably go through the pot and enter the ground, but the main mass of them could be taken up without disturbance,

TREE FERN

For half-shady sheltered dells, in warmer and milder districts, during the summer months.

and then it could if necessary be shifted into a larger pot or small tub.

***Datisca cannabina.**—A distinct and gracefully-habited herbaceous plant from 4 ft. to 6 ft. high. The long stems are clothed with large and handsome pinnate leaves, and the yellowish-green inflorescence appears towards the end of summer. The male plant has long been known as a very strong, graceful, and effective herb. The female plant, however, remains green much longer than the male, and when profusely laden with fruit, each shoot droops and the whole plant improves in aspect. It should not be forgotten in any selection of hardy plants of free growth and imposing aspect. From seed will probably be found the best way to raise it, and then one would be pretty sure of securing plants of both sexes.

Dicksonia antarctica.—A very noble evergreen tree-fern, with a stout trunk, which varies considerably in thickness, and attains a height of 30 ft. or more. The fronds, which form a magnificent crown 20 ft. or 30 ft. across, are lance-shaped, much divided, of a shining dark green on the upper surface, and paler underneath, from 6 ft. to 20 ft. long, beautifully arched, and becoming pendulous with age. Perhaps the hardiest of tree-ferns, and therefore most suitable for placing in the open air in summer in sheltered shady dells, from the middle of May to the beginning of October.

***Dicentra (Dielytra) eximia.**—A plant with dense and very graceful foliage, far more so than any other member of the order in cultivation; and valuable as a flowering plant too, as the brightly-coloured flowers remain on for a long time in spring and early summer. It generally grows from 12 ins. to 16 ins. high, forming thick, almost pyramidal, tufts of pale green, glaucescent, deeply-divided leaves, and bearing handsome drooping clusters of large, rose-coloured flowers, often flowering twice in the year. Although not absolutely necessary, it will be all the better to plant it in light soil. Multiplied by division of the tufts in spring. It should be associated with dwarf subjects like *Thalictrum minus*, or be used on the margins of mixed beds of fine-leaved hardy plants.

***Dimorphanthus mandschuricus.**—A magnificent hardy shrub of erect habit, with very large, much-divided, spiny leaves, which very much resemble those of the Angelica-tree of North America, and in this country attaining a height of 6 ft. to 10 ft., which it will probably much exceed when well established in favourable positions. It is certainly the most remarkable fine-foliage shrub that has been introduced into our gardens for years, and is therefore of the highest importance for the subtropical garden. As to its treatment, it seems to thrive with the greatest vigour in a well-drained deep loam, and would grow well in ordinary garden soil. As to position, isolation in some sheltered but sunny spot will show it to great advantage; but it may also be grouped with like subjects, always allowing space for the spread of its great leaves.

***Dipsacus sylvestris.**—A rather singular-looking hardy native plant, 5 ft. to 6 ft. high, with a prickly, leafy, branching stem, and longish opposite leaves joined together at their bases. The flowers are of a pretty purplish rose-colour, and are borne in conical heads 3 ins.

DIMORPHANTHUS MANDSCHURICUS.

Hardy deciduous shrub Section.

or more in length. Suited for the embellishment of rough, uneven ground, and will grow well in almost any kind of soil. Still more desirable than the foregoing is *D. laciniatus*, a native of France, the leaves of which are fringed with silky hairs instead of spines, and which has always whitish flowers. Both these plants are biennials, and are easily raised from seed.

THE DRACÆNAS.

LONG as this noble family has been known in our gardens, we have yet to learn a great deal about its use and beauty. Hitherto only allowed to grace a stove or conservatory now and then, the Dracænas in future will be among the indispensable ornaments of every garden where grace or variety is sought. They are among the very best of those subjects which may be brought from the conservatory or greenhouse in early summer, and placed in the flower-garden till it is time to take them in again to the houses in which they are to pass the winter months. And if it were not necessary to protect them through the winter, it would be almost worth our while to bring them indoors at that season, so graceful are they, and so useful for adding the highest character to our conservatories. The hardier and most coriaceous kinds, like *indivisa* and *Draco*, may be placed out with impunity very far north. The brightly coloured kinds, like *D. terminalis*, have been tried in the open air at Battersea, but not with success. It would be dangerous to try them in the open air much farther north, except in very favourable spots. The better kinds are indicated in the select list of subtropical plants. *D. indivisa* grows well in the open air in the south of England and Ireland.

Dracæna australis.—A fine plant, with a stout, branched stem, from 10 ft. to 40 ft. high, and oblong, lance-shaped, bright green leaves, from 2 ft. to 3 ft. in length and 2 ins. to 4 ins. in breadth, striated with numerous parallel veins. Flowers white, densely crowded, sweet-scented, $\frac{3}{4}$ in. across; but these are rarely produced except in large houses in botanic or other gardens, where there are old or well-established specimens. This is a useful conservatory species, and may be used out of doors in summer, like *D. indivisa*, though it is not quite so graceful. This plant is of very easy culture in ordinary soil.

Dracæna cannæfolia.—A very fine species from New Holland, with a tall stem, and leaves from 20 ins. to 2 ft. long, the sides of which are rolled inwards, so as to form a kind of pipe, of a fine sea-green colour, and supported on stalks a foot or more in length. This is a valuable species for warm greenhouse or conservatory use, and also for placing out of doors in the southern counties, though it thrives best indoors.

Dracæna Draco (*Dragon tree of Teneriffe*).—A large and vigorous species, native of the Canary Islands, where it attains a great height and age; the now perished Dragon-tree at Orotava in Teneriffe having grown, according to Meyen, "70 ft. high and 48 ft. in circumference, with an antiquity which must at least be greater than that of the Pyramids." In this country it is seldom seen more than 10 ft. high. The stem is tree-like, simple or divided at the top, and often, when old, becoming much branched, each branch terminated by a crowded head of lanceolate-linear entire leaves of a glaucous green colour. The flowers form a large terminal panicle, and are individually small and of a greenish-white colour. This plant is more graceful and effective when in a young state, in which it usually occurs in our gardens, than when old and branched. It thrives well in the greenhouse or conservatory, and in the midland and southern counties may be placed in the open air from the end of May to October, while it is a fine object indoors at all seasons. It is of very easy culture in sandy loam, and requires plenty of pot room.

***Dracæna indivisa.**—A very graceful plant, with leaves from 2 ft. to 4 ft. long, and 1 in. to 2 ins. in breadth, tapering to a point, pendent, and dark green. It should not be confounded with the conservatory plant known as *Cordyline indivisa*, which is too tender to succeed well in the open air, and somewhat difficult to grow. This species, on the contrary, is perfectly hardy in the south of England and Ireland. I saw good specimens of it at Bicton a few years ago, and quite recently as far north as Woodstock, in the county Kilkenny, in Ireland; a plant also stood out in a vase for several years in Mr. A. M'Kenzie's garden at Muswell Hill, N. *D. indivisa lineata* is a very fine variety, the leaves of which are much broader than those of the type, measuring sometimes 4 ins. across, and coloured with reddish pink at the sheathing base. Other good varieties are *D. indivisa atro-purpurea*, which has the base of the leaf and the midrib on the under side of a dark purple; and *D. indivisa Veitchii*, in which the habit and size of the leaf are the same as in the species, but in addition it has a sheathing base and the midrib on the under side is of a beautiful deep red. It would be difficult to find a plant more worthy of cultivation than this. Where it does well in the garden or pleasure-ground in the southern parts, it surpasses any Yucca or other hardy plant that I know in respect of distinctness and tropical-looking grace; and, this being the case, there can be little need to plead for it to the many who have gardens in the counties south of London. In all districts it may be placed in the open garden in summer with fine effect, the rim of the pots plunged level with, or a little below, the surface, and the plants either isolated on the turf, in the centre of a bed of flowering plants, or grouped with other fine-leaved subjects. In the conservatory it is one of the most effective and graceful subjects at all seasons, and is of very easy culture in rich sandy loam.

The *Dracænas* are a very numerous family, and many more might be described; but most of the other kinds have not been proved to possess any excellence for the flower-garden, while those enumerated are abundantly sufficient to represent the aspect of this graceful and stately family.

Echeveria metallica.—This is scarcely high enough to be suitable for association with the taller plants, but it is so very distinct in aspect, and has been proved to grow so well in the open air during several unfavourable seasons, that I must not pass it by. I purposely exclude from this book many things sometimes found in lists of "subtropical" plants, but which may be classed most properly with bedding subjects. But this, although not very large, forms an agreeable and distinct object, and is very well calculated for producing a striking effect among dwarf bedding and edging plants. It may be propagated by the leaves, by cuttings, or by seeds, and requires a dry greenhouse-shelf in the winter. Light sandy earth, not of necessity very poor, will suit it best in the open air. It will prove very effective on the margins of beds and groups of the dwarfer foliage-plants, or here and there among hardy succulents, and should be planted out about the middle of May.

***Echinops ruthenicus.**—A hardy ornamental plant from S. Russia, with stems 3 or 4 feet high, much branched in the upper part, and covered with a silvery down. The leaves are deeply toothed and spiny, of a dark green above, white and cottony underneath. The flowers are blue, and borne in almost spherical heads on the tops of the erect branches. The plant flourishes best in a calcareous soil, but will do well in almost any well-drained ground. Easily multiplied by seed, division of the tufts, or by cuttings of the roots in spring. This is the most ornamental of its distinct family, and is highly suitable for grouping with the finer herbaceous plants. It would also look remarkably bold and well if isolated on the turf.

***Elymus arenarius.**—This wild British grass—a strong-rooting and most distinct-looking herb—is capable of adding a striking feature to the garden here and there, and should be quickly introduced into cultivation. Planted a short distance away from the margin of a shrubbery, or on a bank on the grass, and allowed to have its own way in deep soil, it makes a most striking object. In short, it deserves to rank high among really hardy fine grasses, the Pampas and the two Arundos alone surpassing it. I am not quite certain that it is not more useful than the Arundo, being hardy in all parts of these islands. In very good soil it will grow 4 feet high; and as it is for the leaves we should cultivate it, if the flowers are removed they will be no loss. It is found frequently on our shores, but more abundantly in the north than in the south. The variety called *geniculatus*, which has the spike pendulous, is also worthy of culture, and in its case the flowers may prove worth preserving. It may possibly be useful for covert, and is certainly so for rough spots in the pleasure-ground and in semi-wild places.

***Elymus condensatus** (*Bunch grass*).—A vigorous perennial grass from British Columbia, forming a dense, compact, column-like growth, more than 8 ft. in height, covered from the base almost to the top with long arching leaves, and crowned in the flowering season with numerous erect, rigid spikes, each 6½ ins. long, and resembling an elongated ear of wheat in form. It is a very ornamental plant, and may be associated with our largest grasses. A very distinct variety has been raised in the Edinburgh Royal Botanic Gardens, in which the spikes or ears are much shorter and broader than those of the original form. For this the name *Elymus condensatus compactus* has been suggested.

Entelea arborescens.—A small, branching, light-wooded tree, 5 to 10 ft. high, with large, alternate, heart-shaped or three-lobed leaves covered with stellate down, and white flowers, somewhat like those of a small dog-rose, borne in umbels on the ends of a branching panicle. It is peculiar to New Zealand, and is the only species of the genus. This used to grow satisfactorily in the Paris gardens, but I have no experience of it in this country. It will require greenhouse treatment in winter, and is of but secondary importance for open-air culture.

***Epimedium pinnatum.**—A hardy dwarf perennial from Asia Minor, from 8 ins. to 2½ ft. high, forming handsome tufts of long-stalked radical pinnate leaves, and bearing long clusters of yellow flowers. The handsome leaves remain on the plant until the new ones appear in the ensuing spring. It is not a good plan to remove them, as they serve to shelter the buds of the new leaves during the winter, and the plants flower much better when they are allowed to remain. Cool, moist, peaty soil, and a slightly-shaded position, will be found most suitable for this, and the novel appearance of its foliage claims a place for it among the dwarfer plants, groups of fine-foliaged hardy herbaceous subjects, Mahonias, etc.

***Equisetum Telmateia** (*Giant Horse-tail*). — A British plant of very noble port and much grace of character when well-developed, growing from 3 ft. to 6 ft. high in favourable soil and positions. The stem is furnished from top to bottom with spreading whorls of slender, slightly drooping, quadrangular branches; the whole forming a pyramidal outline of very distinct and pleasing effect. It is a highly ornamental subject for planting in the hardy fernery, the artificial bog, shady peat borders, near cascades, or among shrubs growing best in moist hollows in vegetable soil. Multiplied by division.

***Equisetum sylvaticum** is another native Horse-tail

of much dwarfer size, but of the most exquisite grace when grown; the stem standing from 8 to 15 inches high, and well covered with numerous slender, spreading, or deflexed compound branches. Very suitable for rockwork, margins of ornamental water, or any of the positions in which *E. Telmateia* may be grown. It also does well and looks very graceful when grown in pots in a cold frame. Multiplied by division.

***Erianthus Ravennæ.**—A highly ornamental grass from S. Europe, somewhat like the Pampas grass in habit, but smaller in size, and frequently having violet-tinged leaves. The flowering-stems grow from 5 ft. to 6½ ft. high; but as it only flowers with us in a very warm season, it must be valued for its foliage alone. Its dense and handsome tufts thrive well in light, dry, calcareous soil, in positions with a south aspect. It thrives but poorly on cold soils, and will probably not grow well north of London except in peculiarly favourable positions, and in well-drained free loams. It is fitted for association with such grasses as *Arundo conspicua.* Multiplied by division of the tufts in spring or autumn.

Erianthus Ravennæ.

***Eryngium alpinum.**—A singular-looking plant,

from 2 ft. to nearly 3 ft. high, forming a rather stiff bush, with leathery and very spiny leaves of a sea-green colour, and bearing numerous roundish heads of bluish flowers, the stems beneath them being also of a very handsome blue for some inches down. Suitable for planting in the wilder parts of pleasure-grounds, for isolation, for borders, or grouping with the finest and most distinct subjects.

***Eryngium amethystinum** is not so tall as the preceding, seldom growing more than 2½ ft. high. It is remarkable for the beautiful amethystine bloom which the leaves assume in July, and which they preserve until the approach of frost. It is suitable for the positions recommended for the preceding kind. Various other members of this family are useful in like manner; indeed there is not one of them that is not so, including our own common Sea Holly, *E. maritimum.*

***Erythrina.**—These are very beautiful trees or shrubs, pretty generally distributed through the tropics of both hemispheres. Some attain great dimensions, while others are dwarf bushes with woody rootstocks, and a few have the stems and leaf-stalks beset with prickles. The leaves are trifoliate, with long stalks, and the leaflets oval, lance-shaped, or triangular. Many of the species produce beautiful large pea-flowers, usually of a blood-red or scarlet colour, in terminal racemes. The varieties of these have proved very hardy and useful in the summer garden, flowering freely, and showing considerable beauty of foliage. Two round beds, each of 9 ft., and one oblong bed, 42 ft. by 5 ft., including *E. ornata*, *Marie Belanger*, *laurifolia*, *crista-galli*, *profusa*, *Madame Belanger*, *ruberrima*, *Hendersoni*, stood out last winter uninjured in Battersea Park; and, as many people know, the common old *Erythrina crista-galli* will thrive for years against a warm south wall in a warm soil, if protected about the root in winter.

Eucalyptus.—Handsome Australian trees and shrubs, of which there are a vast number of species, many growing to an immense height. The leaves are of a thick leathery texture, always quite entire, and very variable in shape. In young plants they are opposite, heart-shaped, pointed, and covered with a glaucous bloom; as they grow older, they become alternate and sickle-shaped, the stalks acquiring a peculiar twist, so that the leaves present their edges to the branches. The most hardy kinds are *E. globulus* and *E. Gunni.* Other kinds, however, will no doubt be found sufficiently hardy. These are most likely to be attractive in the south of England and Ireland, where a few of the species will be found to thrive in the open air, as the peculiarly distinct and graceful habit of the trees is not observed till they are 12 ft. or more above the ground. Nevertheless some may grow them for the aspect they present after a single year's growth in the open air about London, in which case they should be put out about the middle of May. I was very much struck with their graceful and singular appearance in California, where they are being planted in great variety.

Farfugium grande.—A very vigorous-growing perennial, with thick fleshy stems, from 1 ft. to nearly 2 ft. high, and broad, cartilaginous, almost heart-shaped leaves, of a light-green colour, variously streaked, and spotted with yellow in one variety, and with white and rose-colour in another. It flourishes best in free, substantial, moist soil which contains a large proportion of vegetable mould, and in a half-shady position. During the heats of summer it will require frequent watering. At the approach of winter it should be removed to the conservatory or cool greenhouse, except in the southern and milder districts, where it survives an ordinary winter. In the colder parts of the country it is scarcely worth planting out, it grows so slowly; but where it thrives it is very ornamental in borders, isolated, or near the margin of beds. Multiplied by division in spring; the offsets to be potted and kept in the propagating-house or in a frame until they are well rooted.

Ferdinanda eminens.—This is one of the tallest and noblest subtropical plants, growing well in the southern and midland counties when it is supplied with rich soil and abundant moisture. It is also very much the better for being sheltered. Where the soil is rich, deep, and humid, and the position warm, it attains large dimensions, sometimes growing over 12 ft. high, and suspending pairs of immense opposite leaves. It will in all cases form a capital companion to the Castor-oil plant, and, though it may not be grown with ease in all parts, it should be in every collection. It requires to be planted out, in a young state, about the middle of May, and grows freely from cuttings. Greenhouse treatment will do in winter. It is better to keep a stock in pots through the summer to afford cuttings, though the old ones may be used for that purpose.

* FERULAS.

I WISH it were not necessary to write in praise of such very fine plants as these, so noble in aspect and beautiful in leaf. If 2000 kinds of herbaceous plants are grown, the first things that show clearly above the ground in the very dawn of spring (even in January) are their deep-green and most elegant leaves. In good garden soil they look like masses of *Leptopteris superba*, that most exquisite of ferns. Their chief charm will probably be found to consist in their furnishing masses of the freshest green and highest grace in early spring. The leaf is apt to lose some of its beauty and fade away early in autumn, but this may to some extent be retarded by cutting out the flower-bearing shoots the moment they appear. Not that these are ugly; for, on the contrary, the plants are fine and striking when in flower. It is indispensable that the Ferulas, like some other hardy foliage-plants, be planted permanently and well at first, as it is only when they are thoroughly established that you get their full effect. At a first view, the best way to treat them would appear to be so to arrange them that they would be succeeded by things that flower in autumn, and only begin their rich growth in early summer; but it will be equally wise to plant them near the margin of a shrubbery, or wherever it is desired to have a diversified and bold type of vegetation. We may look forward to the day when a far greater variety of form will be seen in English gardens than is at present observable, and these Ferulas are thoroughly well worth growing for their superb spring and early summer effect. The best species are *F. com-*

FERULA COMMUNIS.

Graceful herbaceous Type; dying down towards the end of summer, and therefore most desirable for isolation near the margins of shrubberies, etc.

munis and *tingitana*. Probably a few others, including *F. glauca*, *neapolitana*, *nodiflora*, *asparagifolia*, *Ferulago*, and *persica*, may with advantage be added where much variety is sought, but the effect of any of the first three cannot be surpassed. Among the "aspects of vegetation" which we may enjoy in these cold climes, nothing equals that of their grand leaves, pushing up with the snowdrop. In semi-wild spots, where spring flowers abound, it will prove a most tasteful and satisfactory plan to drop a Ferula here and there in a sunny spot, and leave it to nature and its own good constitution afterwards. In general aspect these plants are much alike; it is better, however, to describe some of them individually.

***Ferula asparagifolia.**—An ornamental perennial, 4 ft. or 5 ft. high, with very graceful and finely-cut leaves, the radical ones 1 ft. to 2 ft. long (including the leaf-stalk), repeatedly subdivided: the divisions very narrow, linear, pointed, and set with hairs; the upper stem-leaves are reduced to short sheaths, the lowest of which bear a short pinnate limb.

***Ferula communis.**—A very fine and striking hardy perennial, growing from 6 ft. to 10 ft. or more high, with much-divided, spreading, shining green leaves, repeatedly subdivided into linear, flaccid segments; the lower leaves spreading more than 2 ft. each way; the sheaths of the upper leaves very large.

***Ferula Ferulago.**—A very ornamental kind, with striped stems, 6 ft. to 8 ft. high, and much-divided leaves, with divided spreading leaflets, which are nearly as broad as those of *F. tingitana*, but longer, and of a darker green.

***Ferula glauca.**—A valuable and imposing foliage-plant, 4 ft. to 6 ft. or more high, with very much divided leaves, shining above, glaucous beneath, cut into long, linear, flat segments. The stalks of the upper leaves are widened above.

***Ferula persica.**—A hardy perennial from Persia and the Caucasus, with a glaucous stem from 3 ft. to 6 ft. high, tapering gradually upwards, and very handsome, much-divided leaves, with rather distant lance-shaped leaflets, widening and toothed or cut at the apex. Flowers in umbels, without any general or partial involucrum. This plant is easily distinguished by its strong asafœtida odour, and requires warm sandy soil.

***Ferula tingitana.**—A very noble plant, from 6 ft. to 8 ft. high, with a stout stem and very glistening leaves, which are broader than those of any other species, and repeatedly subdivided into oblong or lance-shaped deeply-toothed segments.

Ficus Chauvieri.—A noble species, with a faultless habit, which does well in the open air, and is the best kind after *F. elastica.* The leaves are oval-obtuse in outline, of a very dark glistening green, with pale-yellow veins, and usually have one or more large undulations on the margin. Useful for the same purposes as the following species.

Ficus elastica (*India-rubber Plant*).—This is one of those valuable leathery-leaved things that are useful in hothouse, drawing-room, or flower-garden. It not only exists in the open air in summer in good health, but makes a good growth under the influence of our weak northern sun. Never assuming the imposing proportions of other plants mentioned herein, it is best adapted for select mixed groups, and, in small gardens, as isolated specimens amongst low bedding plants. It will best enjoy stove treatment in winter, and is propagated from cuttings. It should be put out at the end of May. In all cases it is better to use plants with single stems.

Ficus elastica.

***Fuchsia.** — The Fuchsia, one of the most beautiful ornaments of the garden when well grown, is comparatively rarely seen in our flower-gardens. It is to be regretted that this is the case, for assuredly there is nothing in cultivation more calculated to improve the aspect of things therein. Not showy in mass of flower, like many things common enough now, it is of the highest order of beauty; while the drooping habit of the shoots of most kinds gives the plant a grace which is valuable indeed, and which no flower-garden should be without. Even in dwarf lines, where this drooping tendency is not seen to such advantage, or, it may be, presents a disadvantage, the Fuchsia is very valuable; but it is when we use plants with rather tall stems or pyramids that the full beauty of the Fuschia as a flower-garden plant is seen. And the right way to manage them is to make them as far as possible produce *all their growth in the open air.* That is the secret: start them, nurture them, and make them full of leaves and strong young growth in the spring, so as to go out strong, and most likely you will find them very disappointing indeed; but keep them back and do not let them burst forth into leaf until put in the open air in May, and they will then go on and retain all the strength they gather, suspending quantities of graceful blossoms until the leaves have deserted the trees, when they should be taken up and put in a dry cave, cellar, or shed for the winter. In a cool position of that kind it would not be difficult to "keep them back" in spring. And supposing they seemed inclined to push forth too much before the time had quite arrived when it would be convenient or desirable to put them in the flower-garden, there should be no difficulty in placing them in some quiet, sheltered nook, where they might receive more protection than in the flower-garden proper, and yet have full opportunity to make growth in the open air—the great point to be attained. The freest and hardiest kinds should be chosen for this purpose. In many places refuse plants may be turned to good account in this way. Given a lot of specimen Fuchsias—arrived, perhaps, at that stage when they must be parted with to make way for younger plants and newer kinds—nothing is simpler than to make of these standards for the flower-garden, by cutting away the lower and middle side shoots, and leaving the head to form a standard. Their exceptional grace when placed among fine foliage-plants induces me to allude to them here.

***Funkia Sieboldiana.**—A Japanese plant, remarkable for the elegance of its leaves, which are large, broadly heart-shaped, of a greyish-green colour, slightly undulating, and finely marked with regular lines of prominent ribs. The flowers are of a light lilac or bluish colour, and are borne in a drooping unilateral cluster at the extremity of a leafless stem 1 ft. or 16 ins. high. Thrives best in a light, cool, sandy soil in a sheltered half-shady position. It will also thrive well in peat. Multiplied by division of the tufts in autumn, once in three or four years. Useful among the dwarfer herbaceous plants, etc., and occasionally as a groundwork in beds of shrubs with fine foliage.

***Galega officinalis.**—A handsome, hardy, and vigorous-growing plant, 3 ft. to 5 ft. high, forming graceful tufts of pinnate leaves, and flowering abundantly and for a long time ; the flowers are of a pale blue, in long, dense clusters or spikes. Although it will grow in almost any soil, it does best in a deep, free, moist, sandy clay. This not uncommon herbaceous plant is alluded to here in consequence of its graceful leaves, which fit it for taking a part in the groups of handsome hardy subjects so often suggested in this book.

Geranium anemonæfolium.—A handsome perennial from Madeira, with a simple, woody, erect stem 1 ft. to 14 ins. in height, covered with dry scales. The leaves, which are of a bright green, smooth, and very much divided, are chiefly collected at the base of the plant, from which they extend horizontally on stalks from 20 ins. to 2 ft. long. The flowers are very numerous, and of a lilac rose-colour. It is a highly ornamental plant both in foliage and flower, and may be used with good effect grouped with comparatively dwarf kinds, or occasionally as an edging to tall subjects. It is best raised in frames and put out early in May. Multiplied by seed, which it yields freely.

***Gynerium argenteum** (*Pampas grass*).—This fine plant is so well known that there is no excuse for naming it here, except the opportunity to say a few words as to the splendid use we may make of it in the branch of gardening we are now discussing. It deserves as much attention as any plant in cultivation, and yet how rarely is any thorough preparation made for its perfect development. What is there growing in garden or in wild more nobly distinct and beautiful than the great silvery plumes of this plant waving in the autumnal gusts—the burial plumes as it were of our summer too early dead? What tender plant so effective as this in giving a new aspect of vegetation to our gardens, if it be tastefully placed and well grown? Long before it flowers it possesses more merit for its foliage and habit than scores of things cultivated indoors for their effect—Dasylirions, etc., for example—and it would be well worthy of being extensively used if one of its silken-crested wands were never put forth in autumn. It is not enough to place it in out-of-the-way spots, but the general scene of every garden and pleasure-ground should be influenced by it. It should be planted even far more extensively than it is at present, and given very deep and good soil either natural or made. The

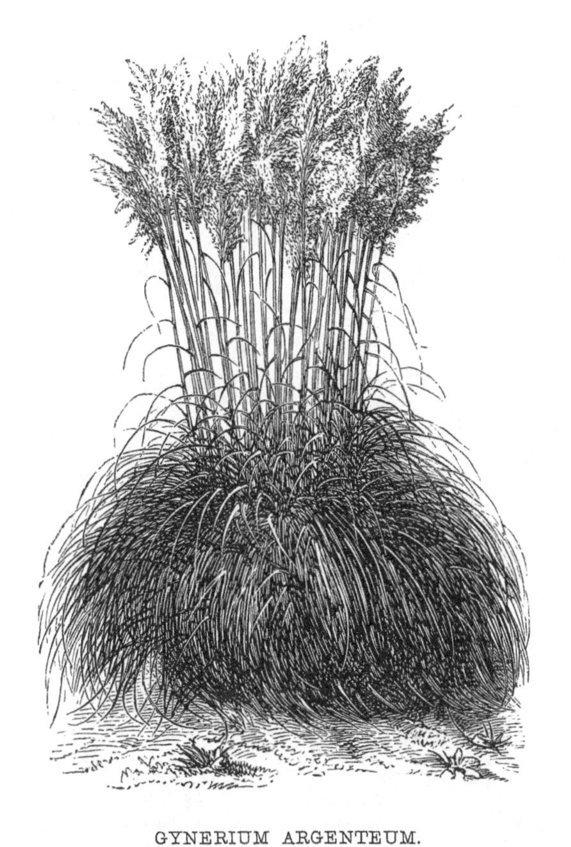

GYNERIUM ARGENTEUM.

Giant hardy evergreen ornamental Grass.

soils of very many gardens are insufficient to give it the highest degree of strength and vigour, and no plant better repays for a thorough preparation, which ought to be the more freely given when it is considered that one preparation suffices for many years. If convenient, give it a somewhat sheltered position in the flower-garden, so as to prevent as much as possible that ceaseless searing away of the foliage which occurs wherever the plant is much exposed to the breeze. We rarely see such fine specimens as in quiet nooks where it is pretty well sheltered by the surrounding vegetation. It is very striking to come upon noble specimens in such quiet green nooks; but, as before hinted, to leave such a magnificent plant out of the flower-garden proper is a decided mistake. Seed and division.

***Gunnera scabra.**—Mr. Darwin met with this in a region where the vegetation is so luxuriant that the branches of the trees extend over the sea, somewhat like those of a shrubbery of evergreens over a gravel walk. "I one day noticed growing on the sandstone cliffs some very fine plants of the Panke (*Gunnera scabra*), which somewhat resembles the rhubarb on a gigantic scale. The inhabitants eat the stalks, which are sub-acid, and tan leather with the roots, and prepare a black dye from them. The leaf is nearly circular, but deeply indented on its margin. I measured one which was nearly 8 ft. in diameter, and therefore no less than 24 ft. in circumference! The stalk is rather more than a yard high, and each plant sends out four or five of these enormous leaves, presenting altogether a very noble appearance." Of a spot in the same neighbourhood he says: "The forest was so impenetrable that no one who has not beheld it can imagine so entangled a mass of dying and dead trunks. I am sure that often for more than ten minutes together our feet never touched the ground, and we were frequently ten or fifteen feet above it; so that the seamen, as a joke, called out the soundings!" Yet I have grown this plant to great size in a cold British bog. Mr. Darwin does not speak of the inflorescence, which is more remarkable than the leaves. The little flowers and seeds are seated densely on conical fleshy masses a few inches long, and these in their turn being seated as densely as they can be packed on a thick stem, the whole has the appearance of a compound cone a couple of feet high (on strong plants), very heavy, and perhaps the oddest-looking thing ever seen in the way of fructification. This great spike springs from the root itself, the leaves also springing from the root, as in the case of the rhubarbs. I had two plants in a wet peat bog—one in deep rich soil, with the crown well raised above the level, and the whole protected under a couple of barrowloads of leaf mould; the other left exposed, and not allowed any particularly good soil. Both plants survived the severest winters, but the protected and well-fed one grew much the larger. The leaves of the larger plant used sometimes to grow 4 ft. in diameter, the texture being of extraordinary thickness and rugosity. I have, however, in the Royal Gardens at Kew, seen it grown to a larger size than that. The bottom there is the reverse of bog, while the situation is warmer and more sheltered than where I grew it. But the Kew people met its wants very cleverly, by building a little bank of turf around it, so

GUNNERA SCABRA.

Hardy herbaceous Section; attaining huge dimensions in deep, rich, moist soil.

as to admit of its absorbing a thorough dose of water now and then, while in winter it was protected with dry leaves and a piece of tarpaulin. Similar protection, plenty of water in summer, and a warm and sheltered position, are all that are necessary for success with this very striking subject. It is not difficult to obtain, and may be raised from seed, though that is a slow way. It should be planted in some isolated spot, and not as a rule in the "flower-garden proper," as it must not be disturbed after being well planted, and would associate badly with the ordinary occupants of the parterre. The plant cannot have too much sun or warmth, but makes little progress if its huge leaves are torn by storms. In places with any diversity of surface it will be easy to select a spot well open to the sun and yet sheltered by surrounding objects (shrubs, clumps, etc.). The figure shows more the habit of the plant than the character of its huge compound fleshy spike, on which the small seeds are thickly scattered.

***Gymnocladus canadensis** (*Kentucky Coffee-tree*).—A remarkable hardy forest tree, which, kept in a young state, will furnish a fine head of foliage. It has twice-divided leaves, the leaflets of which stand vertically. On young trees the leaves are more than 3 ft. long, and on subjects confined to a single stem and cut down every year they would prove even larger than that. It grows wild in Canada, but more abundantly and larger in Kentucky and Tennessee, where it used to be employed as a substitute for coffee—hence the common name. It loves a rich deep soil, and is propagated by seeds and cuttings of the roots. It will prove very suitable for planting in similar positions to those recommended for the Ailantus.

Hedychium Gardnerianum.—A stove herbaceous plant from the East Indies, growing from 3½ ft. to 6½ ft. high, with broadly lance-shaped clasping leaves in two rows, and bearing, in autumn, lemon-coloured flowers with a strong Jonquille perfume, in terminal spikes nearly a foot long. Another species, *H. aurantiacum*, has handsome orange flowers. These handsome plants may have been seen flowering in the open air in the gardens at Battersea Park, very pleasing and effective in shady nooks—not planted out but kept in the pots in which they were grown. They should be put out about the end of May.

***Helianthus annuus** (*Sun-flower*).—This common and familiar annual plant may be usefully employed in adorning the rougher parts of pleasure-grounds in warm positions, where it will thrive to best advantage. Although very hardy and capable of being grown in almost any soils, it prefers those that are substantial and free.

***Helianthus orgyalis.**—A very distinct-looking hardy plant, growing 6 ft. or 8 ft. high. Its distinctiveness arises from the fact that the leaves are recurved in a peculiarly graceful manner. At the top of the shoots, indeed, their aspect is most striking, from their springing up in great profusion and then bending gracefully down. It will form a capital subject for groups of fine-leaved, hardy plants, or for isolation by wood-walks, etc. As it is apt to come up rather thickly, the cultivator will act judiciously by thinning out the shoots when very young, so that those which remain may prove stronger and better furnished with leaves. N. America.

***Hemerocallis fulva.**—This is one of those fine old plants formerly grown in almost every garden, but which

HERACLEUM

Coarse herbaceous Type; foliage perishing late in summer.

has latterly sunk into unmerited neglect, as from its luxuriance, the height of its stem (nearly 4 ft.), and its fine, large, tawny-orange flowers, it well deserves a place among vigorous and striking subjects. It is thoroughly hardy and will thrive in almost any soil, but comes to greatest perfection in that which is deep, substantial, and moist, and in almost any position. Multiplied by dividing the tufts once in three or four years in autumn when the leaves are withered, or in spring. There is a variety of this species with double flowers, and one with variegated leaves, both of which are somewhat tender and require well-drained soil and a sheltered position.

Other species in cultivation are *H. flava*, *H. disticha*, *H. graminea*, and *H. Dumortieri*. All these plants are desirable associates for the hardy fine-leaved plants.

***Heracleums** (*Cow-parsnips*).—No perennials rival these plants in size, and few in distinctness of appearance. The shape and width of their leaves, the height of their stems, and especially the great size of their umbels, produce an effect of a very striking character. Planted singly on slopes in the rougher parts of pleasure-grounds, on or about ruins, rough wild declivities, or by pieces of water or streams, they are seen to excellent advantage; their enormous leaves forming great tufts of vegetation, sometimes nearly 13 ft. in diameter. The period of their greatest vigour and beauty is from May to the end of July; and this should be distinctly borne in mind when arranging them, as, once the great leaves fade away with the heats of early August, they become very ragged, and soon disappear altogether. They delight in a moist, deep, clay soil, but will grow vigorously in almost any kind of cool soil, and even on rubbish-heaps. They are easily multiplied, either from seed or by division of the tufts early in spring, or after the ripening of the seed. They usually sow themselves freely.

***Heracleum eminens.**—A peculiarly distinct species, easily known by the blunt or roundish lobes of its trifoliate leaves, which also possess the advantage of lasting longer than the leaves of other kinds. They are of a thick texture, and finely covered with velvety down, which gives them a slightly glaucous appearance, whereby they may also be distinguished. I have only seen the plant in a young state in Paris gardens, and cannot say what it is like when fully grown; but it is probably dwarfer in stature and more compact in habit than most of the other kinds, and, as the leaves last so much longer in perfection, it is suited for grouping among choicer subjects. For isolation on the grass, after the fashion of the Acanthuses, etc., it is particularly well suited. Seed or division.

***Heracleum flavescens.**—A vigorous species, 5 ft. to 6 ft. high, with deeply-furrowed rough stems, and convex green leaves, with ovate or oblong serrate divisions. Flowers yellowish, in large umbels. Suitable for banks of rivers or ponds, where effective foliage is desired. S. Europe. Seed.

***Heracleum persicum.**—A vigorous perennial, of rapid growth and imposing dimensions, attaining a height of more than 6½ ft., and bearing whitish flowers in umbels, the upper one of which is more than 16 ins. across. The stem does not branch much, and the leaves are chiefly radical and very large. Persia.

***Heracleum pubescens.**—This species bears yellowish-white flowers, in umbels about a foot in diameter, and grows to a height of from 6 ft. to 9 ft. or more. The leaves are very large, roughish on the upper side, and much divided into elliptical segments, which are pointed and stand close together. Crimea.

***Heracleum villosum.**—A species distinguished by its very downy and fringed fruit, with robust stems, 6½ ft. high or more. Leaves broad, much divided, the segments coarsely toothed, covered with a whitish down on the under side. Flowers in large umbels. Caucasus.

***Heracleum Wilhelmsii.**—The leaves of this species are very large, downy, and wrinkled, much divided, with lance-shaped, acute segments distant from each other. The stem is more than 6½ ft. high, and bears whitish flowers, in umbels from 10 ins. to 12 ins. in diameter. Siberia.

Several other species are in cultivation, the principal of which is *H. platytænium*, a biennial of very large dimensions, and remarkably ornamental in habit and foliage.

Humea elegans.—This well-known plant may be very agreeably associated with many of the subjects mentioned in this book, and is a graceful object in the centres of beds, etc. To be put out early in May. New Holland.

***Hydrangea quercifolia.**—This is quite distinct in aspect from the common Hydrangeas, and sufficiently striking to deserve a place where variety of form is sought. The leaves are, as the name indicates, somewhat lobed and oak-like in outline. It is best adapted to form a bush, its natural habit, and is best used singly. I, however, have never tried to train young plants of it with a single stem; it might be worth the trial. It is a native of N. America. A rich and somewhat moist soil, with shelter, will be found to suit it best. Well adapted for isolation.

***Inula Helenium.**—A vigorous-growing British plant, about 3 ft. or 4 ft. high, with a stout stem, branching in the upper part, and large, oblong-oval leaves, of a delicate green colour. The flowers are yellow, and are borne in loose clusters. Well suited for planting along with other large-leaved plants, or as isolated specimens on rough slopes, or wild places, in free, moist, good soil. Multiplied by division in spring or autumn, or by seed.

***Jubæa spectabilis.**—A very handsome, hardy S. American palm, with a short, arboreous, smooth stem, which sometimes attains a height of nearly 40 ft., and spreading pinnate leaves, of a full, deep-green colour, and from 6 ft. to 12 ft. long, the leaflets being from 1 ft. to 1½ ft. long and about an inch wide, springing in pairs from nearly the same spot, and standing out in different directions. The leaf-stalks are very thick at the base, where they are enclosed in a dense mass of rough, brown fibres, which grow upon their lower edges. The soil for this plant should be a mixture of two parts of rich loam and two parts composed of peat, leaf-mould, and sand. This exists in the open air throughout the winter, near London, but not in such a condition as to encourage many to try it in this way. Grown in tubs in the conservatory in winter, and placed in the open air in summer, it will prove very satisfactory for association with the hardier palms.

***Juncus effusus spiralis.**—A very singular-looking

rush, forming spreading tufts of leaves, which, instead of growing straight, like those of other kinds, are curiously twisted in a regular corkscrew form. From its very unusual appearance it is well worthy of cultivation, and may be planted with advantage on the margins of pieces of water, near cascades, etc., or in the artificial bog. It is easily multiplied by division of the tufts.

***Kochia scoparia.**—An annual plant of the Goosefoot family, forming a neat, compact, pointed bush, from 3 to 5 ft. in height. The flowers are green and insignificant, but the graceful habit of the plant renders it valuable and effective, placed either singly or in groups, on the slopes of pleasure-grounds, especially from July to September—the time of its full development. It should be sown in April and May, in a hotbed, and afterwards planted out in beds or borders. Southern Europe.

***Kœlreuteria paniculata.**—A small-sized, hardy, deciduous tree, from N. China, from 15 ft. to 19 ft. high, and handsome both in foliage and flower. The leaves are pinnate, and of a dark, glistening-green colour; the leaflets ovate, and coarsely toothed. Flowers yellow, in terminal, spreading clusters, and succeeded by large, reddish, pendent, bladdery capsules, which render the tree conspicuous till late in the autumn. It does best in rich, moist soil. When planted in a dry and arid position, the leaves are never so large or glistening, and soon turn yellow and fall. It is one of the trees which may, as advised for the Ailantus, be kept in a small state by cutting them down annually, and will, thus treated, furnish a fine effect. Easily multiplied by seed, layers, and cuttings.

Latania borbonica.—A well-known, hardy, and

favourite palm, attaining a height of 25 ft., with large, fan-shaped leaves, over 5 ft. broad, of a cheerful green colour, and with pendent marginal segments. The leaf-stalks are over 4 ft. or 5 ft. long, and are armed at the edges for half their length with short reflexed spines. May be placed in the open air about London and southward in sunny dells in summer, and is a fine warm-conservatory or stove ornament in winter.

***Lavatera arborea.**—This plant, which has the appearance of a small tree, attains in the southern counties the height of nearly 10 ft. The stem is simple for some distance, and then branches into a broad, compact, roundish, and very leafy head. It may be used to adorn warm and sheltered parts of pleasure-grounds and rough places. In rich well-drained beds it would prove a worthy companion for the Ricinus and the Cannas. It is most at home on dry soils, but during the summer months thrives on all. When planted in the shelter of a south wall it has been known to live for several years and to have almost woody stems from 2 ins. to 4 ins. in diameter at the base. Italy.

***Ligularia macrophylla** (*Large-leaved L.*). — A vigorous perennial, with an erect stem nearly 3½ ft. high, and very large, glaucous, erect, long-stalked, oval leaves. The flowers are yellow, and are borne in a dense long spike at the end of the stem. The most suitable soil for this plant is that which is free, moist, and somewhat peaty. Multiplied by careful division in autumn or in spring. It is a useful subject for grouping with fine-leaved herbaceous plants, but will seldom command a place in the select flower-garden. Caucasus.

***Malva crispa.**—A vigorous-growing annual, 3 to 6½ ft., or more, in height, growing in an erect, pyramidal bush of densely-crowded, broad leaves, with a very undulating curled or frizzled margin. The flowers are small, white, and inconspicuous. Bushes of this are pretty in groups, beds, or borders. It may be sown in cool frames and put out early in May, by which means strong plants may be obtained early in the season.

Malva crispa.

***Martynia lutea.**—A very pretty annual from Brazil, about 1½ ft. high, with roundish leaves and handsome yellow flowers, collected in cylindrical clusters at the angles of the principal branches. It requires a light, rich, cool soil, a warm position, and frequent watering in summer. Its large leaves and ornamental bloom make it a desirable subject for beds, groups, and borders. Increased by seed.

***Megaseas.** — Some parts of pleasure-grounds for which there may be a difficulty in finding any subject that will either thrive in such places, or that is not too commonplace to be thought of, may be very advantageously occupied by a few specimens of *Megasea cordifolia* and *M. crassifolia*, the large, leathery, persistent leaves of which will, in such positions, preserve their freshness through the winter, and in the earliest days of spring be surmounted by dense clusters of very pleasing rose-coloured flowers. These plants are very hardy perennials, and will require hardly any care after planting them. They are well suited for embellishing cascades and rough rockwork, are not fastidious as to soil, and are easily multiplied by division of the tufts.

Melanoselinum decipiens. — An umbelliferous shrub, from Madeira, with a round simple stem, bare below, and large spreading compound leaves with oval toothed segments and sheathing leaf-stalks. Flowers white, borne in umbels. Should be planted out in May. A useful subject for isolation on grass-plats, etc., and requiring greenhouse or warm-frame treatment in winter. Young plants are to be preferred for ornamental purposes. Multiplied by seed.

Melia Azedarach (*Pride of India*).—A very ornamental tree, but with us usually seen in a small state. It is a native of India, but is now naturalised in various parts of S. Europe. It usually grows from 13 ft. to 17 ft. high, and bears its smooth dark-green pinnate leaves chiefly clustered together at the ends of the branches. The flowers are of a lilac colour, and emit an agreeable perfume: they are produced in large bunches which issue from the axils of the leaves. The fruit is considered poisonous. Healthy young plants kept to a single stem, and cut down to within a foot or two of the base annually, form graceful objects in the ornamental garden, throwing up an erect stout stem regularly clothed with leaves twice and thrice divided, from 16 ins. to more than 2 ft. long and from 1 ft. to nearly 18 ins. across. Plant out in early summer. A substantial, well-manured soil, a shady position, and frequent waterings in summer are necessary for

MELIANTHUS MAJOR.

Herbaceous Type: best treated as a perennial, and protected at the roots in winter.

the development of the ornamental qualities of the Melia. In winter it requires the shelter of a greenhouse with us. Multiplied by seed or layers.

***Melianthus major.**—This is usually treated as a greenhouse plant, and is sometimes put out of doors in summer. So grown, however, the full beauty of the plant has not time to develope; and much the better way is to treat it as a half-hardy subject, putting it out in some sunny and sheltered spot, where the roots will not suffer from wet in winter. The shoots will be cut down by frost, but the root will live and push up strong stems in spring, forming by midsummer a bush about 3 ft. high, very distinct and beautiful, with large, pinnate, light sea-green leaves, which will not suffer from rain storms. I have grown it in this way to a much more presentable condition than it ever assumes indoors, where it is usually drawn too much. I used to protect the roots in winter by placing leaves over them, and then covering all with a handlight, but have seen the plant survive without this precaution. It is, however, best to make quite sure by using protection, except where the soil and climate are particularly favourable. *Melianthus minor* has the habit of *M. major*, but is smaller in every respect and flowers later. Its inflorescence also is different, the flowers being of a yellowish red, and growing in short clusters. Cape of Good Hope.

***Meum athamanticum.**—One of the most elegant and gracefully-cut plants in an order second to no other for these qualities. The leaves are divided so finely as to appear as if they were made of vegetable hair, and the plant is dwarf and neat in habit, from 6 inches to 1 foot high; easy of growth in ordinary soils, and perfectly hardy and perennial. It is probable that in dry seasons it might "go off" too soon for association with autumn-flowering plants, but for rockwork, borders, or mixed arrangements of any sort it is invaluable. A British plant, easily increased by division.

***Molopospermum cicutarium.**—A very ornamental umbelliferous plant, 5 ft. or more in height, with large, deeply-divided leaves of a lively green colour, forming a dense irregular bush. The flowers, which are insignificant and of a yellowish-white colour, are borne in small roundish umbels. There is a deep green and fern-like beauty displayed profusely by some of the umbelliferous family, but I have rarely met with one so remarkably attractive as this species. Many of the class, while very elegant, perish quickly, get shabby indeed by the end of June, and are therefore out of place in the tasteful flower-garden; but this is firm in character, of a fine rich green, stout yet spreading in habit, growing more than a yard high, and making altogether a most pleasing bush. It is perfectly hardy, and easily increased by seed or division, but rare as yet. Loves a deep moist soil, but will thrive in any good garden soil. It is a fine subject for isolation or grouping with other hardy and graceful-leaved umbelliferous plants. Carniola.

Monstera deliciosa.—This very remarkable-looking plant has been found to bear being placed in the open air with impunity in shady and sheltered spots from the beginning of June till the end of September. Its great perforated leaves look so singular that everybody should grow it who has a stove in which to do so, and it is so

MONSTERA DELICIOSA

Tender Section; stands well in the open air in summer in warm and sheltered spots.

MONTAGNÆA HERACLEIFOLIA (*Polymnia grandis*).

One of the noblest of the tender section, making free growth in the open air during summer.

readily grown and propagated that a plant may soon be spared for placing in the open air during the warmer months. Although, however, it will exist in the open air for a few months in warm and sheltered spots in this country, it is only to be seen to perfection in the stove. Few subjects possess nobler or more singular foliage, or are more worthy of a place for effect alone, while its fruit is edible and produced freely enough when the plant is well-grown. The culture is quite simple—plenty of rich soil, a high temperature, and abundance of moisture. I have seen the plant fruit well in what is called an "intermediate house," but, where convenient, it is much better to place it in a warm stove. Wherever rockwork or any picturesque arrangement is attempted in any of these structures, no nobler plant can be selected for its embellishment. It crawls over such positions with a snake-like stem, and a trail of great leathery leaves perforated in many places, as shown in the plate. Mexico.

Montagnæa heracleifolia (*Polymnia grandis*).—This is second to no other plant for its dignified and finished effect in the flower-garden, forming a handsome shrub with large, opposite, much-divided, and elegantly-lobed leaves, which are often nearly 3 ft. long, presenting very striking and luxuriant masses of foliage. The stem and leaf-stalks are spotted with white, and the leaves when young are covered with a soft white down. Like most large soft-growing things in this way, it is best planted out in a young state, so as to ensure a fresh and unstinted growth. Easily multiplied from cuttings, which plants freely produce if placed in heat in January.

It is best planted out at the end of May, and should be in every collection. Mexico.

***Morina longifolia.**—A hardy perennial of handsome and singular appearance, the large spiny leaves resembling those of certain Thistles, while the long spikes of whorled flowers, which are from 2 to 3 ft. high, are almost identical in form with the inflorescence of many of the Labiate family. It grows well in ordinary well-drained soil, preferring that which is mellow, deep, and moist. Easily multiplied by sowing the seed as soon as it ripens, in light, peaty, sandy soil. In our winters it would be advisable to cover the plants with litter, as it is the dampness and sudden changes of the atmosphere which appear to injure it rather than the cold. Being a fine-flowering plant, as well as remarkable for its leaves, it is excellent for every kind of mixed border, and also for grouping with the smaller and medium-sized perennials that have fine foliage or are singular in appearance. Nepaul.

Morina longifolia.

***Mulgedium alpinum.**—A tall mountain-plant of the Sow-thistle family, with very broad leaves and stems over 3 ft. high. The flowers are very numerous, of a dark-blue colour, resembling those of the Chicory, and are borne erect on the upper parts of the branching stems. A somewhat shaded position and a very moist deep loam are necessary to the vigorous growth of this plant, which will prove effective in the rougher parts of the pleasure-ground. Multiplied by division. Alps.

Mulgedium alpinum.

***Mulgedium Plumieri.**—A vigorous herbaceous perennial, 4 ft. to 6 ft. high, with very large, deeply-cut radical leaves somewhat glaucous underneath, bearing late in summer large spreading terminal corymbs of blue flowers. A good subject for association with strong-growing herbaceous plants in groups, or as isolated tufts by wood-walks, in deep rich soil. S. of France.

Musa Ensete.—The noblest of all the fine-leaved plants yet used in the flower-garden is *Musa Ensete*—the great Abyssinian Banana, discovered by Bruce—the stem of which has been known to attain a diameter of more than 3 ft. at the base, and a height of from 13 ft. to nearly 20 ft. The huge leaves, borne nearly erect, are oblong in shape and of a beautiful bright-green colour, with a very stout deep-red midrib. In the open air they often grow nearly 10 ft. long and nearly 2 ft. broad. The fruit of this kind is not edible, like that of the Banana and Plantain (*Musa paradisiaca* and *M. sapientum*), but the leaves are magnificent, and they stand the rain and storms of our northern climes with little laceration, while all the other kinds of Musa become torn into shreds. It has hitherto been generally grown in stoves; but it is an interesting and, until recently, little known fact, that the finest of all the Banana or Musa tribe is also the hardiest and most easily preserved. When grown for the open air it will of course require to be kept in a house during winter and planted out the first week in June. In any place where there is a large conservatory or winter-garden it will be found most valuable, either for planting therein or for keeping over the winter, as, if merely housed in such a structure during the cold months, it will prove a great ornament among the other plants, while it may be put out in summer, when the attraction is all out of doors. Other kinds of Musa have been tried in the open air in England, but have barely existed, making it clear that they should not be so cultivated in this country. *M. Ensete* is the only species really worth growing in this way. Where the climate is too cold to put it out of doors in summer, it should be grown in all conservatories in which it is desired to establish the noblest type of vegetation. It also stands the drought and heat of a dwelling-house remarkably well, and though, when well developed, it is much too big for any but Brobdingnagian halls, the fact may nevertheless be taken much advantage of by those interested in room-decoration on a large scale. The plant is difficult to obtain as yet, but will, I trust, be soon made abundant by our nurserymen.

In September, 1868, I saw a fine plant of this Musa

MUSA ENSETE

Noblest of tender Section; thriving in the open air in summer, in the warmer districts in summer and in the conservatory in winter.

that had remained in the open ground in Baron Haussmann's garden in the Bois de Boulogne during the preceding winter. It was left in the position in which it grew during the summer of 1867, and in the month of November covered with a little thatched shed, the space about the plant being filled with dry leaves. All the leaves were cut off. In spring the protection was removed and the plant pushed vigorously. It had, when I saw it, 16 leaves, not one of which was torn or lacerated, although it was in an exposed position. It was not more than 5 ft. high, but was more attractive than much larger individuals of the same species, from being so compact and untattered in its foliage. As most people who grow it will have means of keeping it indoors in winter, and as it is so rare, this mode of keeping it is not likely to be taken advantage of with us at present; but that it can and has been so wintered is an interesting fact.

Other species are: *M. paradisiaca* (the Plantain); *M. sapientum* (the Banana); *M. Cavendishii*, a dwarf-growing kind, seldom exceeding 5 ft. or 6 ft. in height, and fruiting freely in a stove; *M. superba*, and *M. zebrina*. These, however, are all properly hothouse kinds, and do not thrive in the open air.

Nelumbium luteum.—This is perhaps second to none of its fine family in beauty. In the Paris Garden a plant remained in a fountain-basin, in a very sheltered position, for several years, flowering yearly. On the 11th of September, 1868, the petals of one of these flowers measured each 6 ins. in length! and consequently the flower fully spread out was more than a foot in diameter. The singular-looking disk within these great fragrant flowers is as interesting as the flower itself, and far more peculiar. The flowers are of a pale yellow, with a single deep stain of rosy crimson at the apex of each petal. The leaves push boldly above the water, are quite round, 12 ins. to 15 ins. in diameter, and of a peculiar bluish-green. If the hand be placed under one of them, so as to slightly raise the outer parts of the large plate-like leaf, a hollow will of course be formed; and if water be poured into this so as to cover exactly the centre of the stem which supports and feeds the leaf, a curious result will be noticed. Bubbles of air will rise rapidly one after another from exactly over the part where the leaf joins its stem. In this particular leaf this spot is marked by being of a much lighter green than the remainder of the leaf. This would be well worth trying in the southern counties on the margin of ponds, etc., where it would be fine associated with *Nuphar advena*, and the like. It should be put out early in May.

Nelumbium speciosum (*Sacred Bean*).—An Egyptian water-plant of very great beauty, with round sea-green leaves from 10 ins. to over a foot in diameter, some of which float on the surface of the water, while others are elevated above it at various heights up to 2½ ft. Above the highest leaves appear the large, solitary, white, rosy-tipped flowers, which are about 10 ins. broad, and are very fragrant. As the plant is somewhat tender, it should be grown in tubs, and placed in rather shallow water (which is more easily warmed by the sun) from the end of May until the end of September in a position

NICOTIANA TABACUM. (*Var. macrophylla*). (8 ft.).

Annual Section; forming handsome specimens within a few months after sowing.

sheltered from sudden changes of the atmosphere. A mixture of good loam, sharp sand, and charcoal seems to answer it best. Multiplied by division of the root-stocks, or by seed. Wherever there is a contrivance for heating the water in a small pond or tank in the open air, these two remarkable plants would be well worth a trial.

Nicotiana Tabacum (*Common Tobacco*). — There are several varieties of this well-known plant, distinguished by the different length and width of their oblong lance-shaped leaves and the lighter or deeper colour of their rosy flowers. A deep, well-drained, light soil, rich in decayed vegetable matter, is essential to this plant, as is also copious watering in summer. As it is so readily raised from seed, and grows luxuriantly in rich soil, I need not say it is a very desirable subject for association with the Castor-oil plant and the like, and especially suited for the many who desire plants of noble habit, but who cannot preserve the tender ones through the winter under glass. It should be raised in a hotbed, and put out in May. Height, 6 ft. or more. *N. macrophylla* is the finest variety.

Nicotiana wigandioides.—A species of recent introduction, with a simple stem 6 ft. or more high, and very large, broad, woolly leaves which grow in a somewhat erect position. If placed in a hothouse at the close of the year, it becomes woody and branching, and assumes an arborescent habit. Requires good substantial, well-manured soil, and copious watering. This is a useful plant for grouping with the medium-sized tender subjects in rich, light, well-drained soils, and should be planted out about the middle of May.

***Nuphar advena.**—A hardy Water-lily from N. America, with broad, oval-heart-shaped leaves, some floating on the surface of the water, and some emerging well above it. The flowers are rather large, of a golden-yellow colour, with scarlet stamens. This plant is as hardy as our native water-lilies, and is therefore adapted for any position in which they will grow. In N. America it occurs chiefly in still or stagnant waters. The bold and large leaves make it peculiarly suitable for placing near the margin of water in the vicinity of groups of picturesque plants.

***Onopordum Acanthium.**—A native biennial plant of bold habit and vigorous growth, with stout, branching stems often more than 5 ft. high, and very large, undulating, spiny leaves, covered with long, whitish, cobweb-like hairs. Flowers purplish, in large, solitary, terminal heads. *O. illyricum* has greener and more deeply-cut leaves, stiffer stems, a more branching habit, and much more spiny leaves and stems. *O. arabicum* grows to the height of nearly 8 ft. with an erect and very slightly branching habit, and has both sides of the leaves, as well as the stems, covered with a white down. All these

Onopordum Acanthium.

species are very hardy, and thrive well in ordinary soil. They are particularly adapted for the rougher and more neglected parts of pleasure-grounds, where they will sow themselves.

***Osmunda regalis.**—This, the largest and most striking of our native ferns (sometimes attaining a height of 8 ft.), is one of the most ornamental subjects that can be grown in certain positions. It should be planted in moist peaty soil in half-shady places, on the banks of streams, the margins of pieces of water, by cascades, etc. It may also be planted in the water with good effect. It has been found to do well exposed to the full sun, when its roots are in a constantly moist, porous, moss-covered soil, in a position sheltered from strong winds. In shady positions it would be found to attain great stature if planted in deep, well-manured soil. The various North American Osmundas should also be associated with this. There are several varieties to be met with in gardens, the handsomest of which perhaps is *O. r. cristata.* Popular and almost universally cultivated as the Osmunda is, however, it is not at all common to see the Royal Fern and several other Osmundas otherwise than in a shabby, or at best in a half-developed, condition. Mr. A. Parsons, of Danesbury, a well-known florist and cultivator of ferns, has overcome this difficulty, and narrates his marked success in the pages of the *Florist and Pomologist.* He formed a very large fernery in an old chalk-pit, and with much success; but, notwithstanding all the care taken of the Osmundas and allied ferns, they were tried for four seasons with no satisfactory results, the roots of the surrounding trees robbing them of both soil and water. "A change was then made: a piece of ground, of irregular shape, large enough to contain about 20 plants, was staked out, and the mould, or, more correctly speaking, the chalk, was removed to the depth of 3 ft.; a bricklayer followed, and put in a floor of three bricks laid on the flat, set in good Portland cement, and over that a layer of plain tiles, the sides being made up to the ground-level with a 4½-inch wall, well built up in the same kind of cement; this made the whole water-tight, and prevented the roots of the surrounding trees from penetrating and robbing the ferns of their moisture. The space was filled up with earth, compounded of good loam, peat, and leaf-mould, in equal proportions, with about one-fifth of good rotten manure added thereto; these ingredients were thoroughly mixed and well trodden in, and then the ferns were planted. In forming this bed, provision was made for the escape of the surplus water, by introducing into the front wall, at about 4 ins. from the bottom, a common 3-in. drain-pipe, which communicated with a small tank, about 3 ft. square, sunk into the chalk, so that all waste water became absorbed. This method proved to be eminently successful, the plants far surpassing in size any I have ever seen under artificial cultivation, and, judging from report, rivalling their growth in their natural habitats. Last season I could boast of *Osmunda regalis* with fronds at least 8 ft. in length, *Osmunda spectabilis* 4½ ft., *Osmunda Claytoniana* 5 ft., *Osmunda cinnamomea* 3 ft., and the beautiful *Osmunda regalis*, var. *cristata*, 3 ft. in length. *Adiantum pedatum* grew from 2 ft. to 3 ft. in height, and others were proportionally fine. The plants were not drawn up by being planted closely together, but were placed at a fair distance apart, and became handsome and noble specimens. Every spring I apply a dressing of about two inches of rotten manure to the surface, and just cover it with mould for the sake of appearance. The artificial swamp is the admiration of all the visitors here. The plants are always in a healthy and vigorous state, and have none of that half-starved appearance so frequently to be seen. The result of my experience induces me to believe that a more liberal treatment would not be found objectionable in the cultivation of many more of our native ferns. I intend making the experiment this season, and may possibly find time to make known what amount of success I may meet with. In concluding my remarks upon what I may term 'growing Osmundas under difficulties,' I would observe that the points to be principally attended to are—(1) a deep water-tight and root-tight tank, the depth of which may, with advantage, be more than in the case I have described; (2) a rich nutritious soil; (3) a reasonable amount of water; and (4) a drain to carry off the surplus."

***Pæonia tenuifolia.**—A charming plant, about 1½ ft. or 2 ft. high, with numerous upright stems almost completely hidden by the dense finely-cut leaves, producing an effect which very much resembles that afforded by the foliage of the Ferulas. The flowers are of a deep crimson, or purplish-red, and about 2½ ins. broad, solitary and erect on the tops of the stems. It does well in almost any soil or position, and might with advantage be now and then used as an isolated specimen on the turf.

***Panicum altissimum.**—A very handsome hardy perennial grass, very like *P. virgatum*, and often confounded with it, although much more elegant in habit. It forms dense, erect tufts from 2 ft. to 6½ ft. high, according to climate, soil, and temperature. The leaves are linear, finely toothed, long, and arching. When in flower the plant presents a very attractive appearance, the inflorescence consisting of very large panicles of slender whorled branchlets covered with numerous small spikelets of flowers, which ultimately assume a tinge of dark chestnut-red.

***Panicum bulbosum.**—A tall and strong species, with a free and beautiful inflorescence. It grows about 5 ft. high, and the flowers are very gracefully spread forth. It forms an elegant plant for the flower-garden in which grace and variety are sought; for dotting about here and there, near the margins of shrubberies, etc.; and for naturalisation.

***Panicum capillare.**—A hardy annual kind, growing in tufts from 16 ins. to 20 ins. high, and very ornamental when in full flower, the tufts being then covered with large, pyramidal panicles, which are borne both at the ends of the stems and in the axils of the stem-leaves. Grows in any soil or position, and sows itself. I noticed it in great abundance in cultivated fields in many parts of Canada and the United States. Well suited for border beds or isolation, being one of the most graceful plants in cultivation. It is commonly grown under the name of *Eragrostis elegans*.

***Panicum virgatum.**—A handsome, bold, hardy species from N. America, growing, in good soil, to a height of nearly 3½ ft. It forms close, compact tufts of leaves, a foot or more long, which, from July to the first frosts, are crowned with very large, dense, branching panicles. The general colour of the plant is a fine lively green, and its graceful habit renders it an admirable subject for the picturesque flower-garden, the pleasure-ground, etc., in isolated tufts. It is also fine for borders. The best mode of multiplying it is by division in the spring, when vegetation is just commencing.

***Papaver bracteatum** (*Great Scarlet Poppy*).—A remarkably vigorously-growing perennial species from Siberia and N. Russia, with simple, rough, hairy stems more than 4 ft. high, each terminated by a single flower 9 ins. broad, and of a bright deep brick-red colour, with a large black spot at the base of each petal, and 2 or 3 leaf-like bracts at the base of each flower. The radical leaves, which are very deeply divided, form a dense broad tuft resting on the ground. A very ornamental plant for the open parts of pleasure-grounds, flourishing in almost any kind of well-drained soil. It will prove most effective as an isolated plant in the rougher parts of the pleasure-ground. A plant of it would also show to great advantage in a group of green-leaved subjects like the Ferulas. Very closely resembling the preceding is the Armenian species *P. orientalis.* It is however smaller in every respect, and the flowers want the distinguishing bracts.

Papyrus antiquorum (*Egyptian Paper-plant*).—A very graceful reed, or rather cyperus, which yielded the material used as paper by the ancient Egyptians. The rootstocks spread horizontally under the mud in places where the plant grows, continuing to throw up stems as they creep along. These stems are triangular and grow from 3 ft. to 8 ft. high; they are quite leafless except at the top, where they bear a large parasol-like tuft of green, gracefully-arching filaments. Shallow parts of pieces of water in a warm position are the most suitable places for this plant. It may, however, also be planted in soil which is kept constantly moist. Multiplied by division of the rootstocks; the pieces to be planted under water, if possible. In gardens south of London this fine plant may be tried in the open air in summer from June till September in warm spots; if not planted in shallow water, or the artificial bog, it should receive a very liberal supply of water in summer.

***Paulownia imperialis.**—A Japanese tree of moderate size, with a large, dense, spreading head, and broad, entire or lobed, opposite leaves covered with hoary down. The flowers are nearly 2 ins. long, in terminal panicles, and of a purplish-violet colour. Young plants, say of a year, or in the first spring of their existence as independent plants, will, if cut down to the ground, make a luxuriant growth during the current year, and indeed prove more effective than the *Ferdinanda eminens.* The stem rises quite vertically and with great vigour, and looks simply a column of noble leaves. Plants growing in an ordinary nursery-quarter were (Sept. 8, 1868) 7 ft. and 8 ft. high from the ground, and the leaves from 20 ins. to 22 ins. in diameter. It was noticeable, too, that those at the top of each shaft were as large as those half-way up and near the base, which is not usually the case; and of course every variation of this kind is valuable, as it helps us to produce variety. Planted in rich ground and treated specially to secure a subtropical effect, greater dimensions than the above may readily be obtained, and older plants when cut down may be expected to produce stronger stems and leaves. They should be cut down every year in winter or spring, and confined to one stem.

***Petasites vulgaris.**—A native plant of vigorous growth and striking appearance, forming dense spreading tufts of enormous radical leaves, usually about 2 ft. or 2½ ft. high. The leaves are heart-shaped in outline, and sometimes 3 ft. in width. A suitable subject for planting in any position where the soil is moist and deep, as on the margins of pieces of water or in damp glades. There is a variety (*P. v. niveus*) which has white, instead of rosy, flowers. This is also quite hardy, but does best in a slightly shaded position in deep clayey or sandy-clay soil.

Phœnix dactylifera (*Date Palm*).—A handsome palm with a rugged stem, and pinnate dark-green leaves from 6 ft. to 12 ft. long; the divisions linear-lance-shaped, very much pointed, and standing out quite straight. Easily increased from seed. Suitable for the greenhouse in winter and the open garden in summer, from the end of May till the beginning of October. Africa and India.

Phormium tenax (*New Zealand Flax*).—A hardy plant, with something of the habit of a large Iris, forming tufts of broad, shining, leathery leaves from 5 ft. to 6½ ft. high, gracefully arching at the top. The flowers, of a lemon colour, are borne in erect loose spikes just above the foliage. Generally with us it will be found to enjoy greenhouse temperature, though in genial places in the south and west of England and Ireland it does very well in the open air. Its best use is for the decoration of the garden in summer, a few specimens well grown and plunged in the grass or the centre of a bed giving a most distinct aspect to the scene. The larger such plants are, the better, of course, will be the effect. The small ones will prove equally useful and effective in vases, to which they will add a grace that vases rarely now possess. It is pre-eminently useful from its being alike good for the house, conservatory, and hall-decoration in winter. Multiplied by division of the tufts in summer, and thriving best in a light deep soil. Wherever indoor decoration on a large scale is practised it is indispensable, and it should be remarked that, unless for vase decoration, it requires to be grown into goodly specimens before affording much effect out of doors; but when grown large in tubs, it is equally grand for the large conservatory and for important positions in the flower-garden. In the extreme south of England and Ireland the New Zealand Flax will thrive in the water as well as on dry land; and where this is the case it may of course be used with fine effect as an aquatic. Doubtless, too, the variegated variety of the New Zealand Flax would be a capital plant to try in the open air in nice situations in the south and west of England and Ireland, where the green plant sometimes does so well. In any case it will do finely out of doors for the summer.

***Phytolacca decandra** (*Virginian Poke*).—A vigorous herbaceous perennial, from 5½ ft. to nearly 10 ft. high, with stems of a reddish hue, very much branched above; the branches, leaf-stalks, veins of the leaves, and flower-stalks being also red. The flowers are numerous, in cylindrical spikes, and are at first white, afterwards changing to a delicate rose-colour. In autumn the leaves change to a uniform reddish tinge, which has a fine effect, contrasted with the numerous pendent purple berries. This is a very hardy plant, requiring hardly any attention and growing in almost any kind of soil. Multiplied either by seed or by division. It forms a very free and vigorous mass of vegetation, and, though perhaps scarcely refined enough in leaf to justify its being recommended for flower-garden use, no plant is more worthy of a place wherever a rich herbaceous vegetation is desired; whether near the rougher approaches of a hardy fernery, open glades near woodland walks, or any like positions. N. America.

***Poa aquatica.**—A stout, rapidly increasing native grass, growing from 4 ft. to 6 ft. high, with broad, flat leaves and the inflorescence in much-branched handsome panicles, sometimes nearly a foot long. It is not uncommon in England and Ireland, mostly occurring in wet ditches, by rivers, and in marshes. It is one of the boldest and handsomest hardy grasses, for planting by the margins of pieces of artificial water or streams, associated with such plants as the Typhas, Acorus, bullrush, great water-dock, etc.

***Poa fertilis.**—Just within the main entrance of the Royal Gardens at Kew a very graceful-looking grass might have been seen isolated on the turf during the past year or two. It is a comparatively dwarf subject, and not at all striking in bloom like the Pampas, but withal very distinct and desirable. It is one of the most elegant grasses, forming dense tufts of long, soft, smooth, slender leaves, which arch outwards and downwards in the most graceful manner on every side, and, in the flowering season, are surmounted by airy, diffuse, purplish or violet-tinged panicles, rising to a height of from 20 ins. to 3 ft., the grassy tufts being usually about half that height. This plant is widely distributed over South-

Poa fertilis.

ern Europe, Northern Asia, and North America, in wet meadows and on low banks of streams. Of all the dwarf perennial grasses it is perhaps the best for isolation on the grass, where its fine dense and graceful tufts of long hair-like leaves and elegant panicles form a quite distinct-looking and ornamental object.

***Polygonatum multiflorum** (*Solomon's Seal*).—This, one of the most graceful of our native plants, is too distinct and pleasing in aspect to be omitted from an enumeration of ornamental subjects. It is best suited for a shady position under trees, or the fringes of shrubberies, or groups of tall and widely-branching plants, where its elegantly arching stems and pretty pendent flowers would attain greatest perfection, and exhibit a very marked contrast to the surrounding types of vegetation. A well-drained, sandy, and peaty soil will be found to suit it best, and it is well adapted for the wild and semi-wild parts of the pleasure-ground. Easily multiplied by division of the rootstocks.

***Polygonum cuspidatum** (*Sieboldi*).—Forms large and noble tufts of lively green, which increase in beauty from year to year. It grows to a height of 3½ ft. to 6 ft. and more, the stems being at first erect and simple, then becoming much branched, the branches arching, and spreading nearly horizontally at the top. The white flowers, which are disposed in clusters forming close panicles, are succeeded by handsome rosy-white fruit. When planted singly, and away from other subjects, its head assumes a rather peculiar and pretty arching character; and therefore it is not quite fit for forming centres or using in groups, so much as for planting singly on the turf, there leaving it to take care of itself and come up year after year. In this way it would be particularly useful in the pleasure-ground or diversified English flower-garden. It is also good for any position in which a bold and distinct type of vegetation is desired, while of course, when we come to have fine groups of hardy "foliage-plants" in our gardens, its use will be much extended. The deeper and better the soil, the finer will its development prove. You cannot make the soil too deep and good if you want the plant to assume a very striking character. It runs very much at the root in all directions horizontally just below the surface of the soil. By cutting away the runners, and thus concentrating the sap in the central stems, tufts have been obtained from 9 ft. to 13 ft. high, and as much across. Japan.

Polymnia grandis. (See *Montagnœa heracleifolia.*)

Polymnia pyramidalis.—A free, bold, and tall plant, with somewhat of a sunflower habit, but withal very fine, and making a tall green growth by the end of July, before many other things used in this way begin to push. The leaves are not so large as those of the other species, and differ in shape, being nearly cordate; but the growth is vigorous, and the habit distinct. It pushes up a narrow pyramidal head of foliage to a height of nearly 10 ft. in Paris gardens, and will be found to do well in the south of England. Easily multiplied in spring by division or cuttings from plants placed in heat in January, or from seed sown in a pot. The species *Uvedalia* and *maculata* are rather coarse herbaceous plants, fitted for rough places in warm positions, and deep, rich soil. New Granada.

***Pontederia cordata.**—One of the handsomest water-plants in cultivation, combining gracefulness of habit and leaf with beauty of flower. It forms thick tufts of almost arrow-shaped, erect, long-stalked leaves from 1½ ft. to more than 2 ft. high, crowned with the handsome blue flower-spikes, which issue from the leaf-stalks just below the base of the leaves. It should be planted in shallow pieces of water. Multiplied by division of the tufts at any season. N. America.

Pothos acaulis.—A noble plant of the Arum family,

with huge simple dark-green leaves, forming a magnificent rosette. It requires stove treatment in winter and spring, and having made its growth and been hardened off under cover, it may be placed out in the open air in sheltered warm places in the southern counties, from the middle of June to the end of September. It, however, sometimes suffers from cold, and is on the whole only likely to be of very partial use. Among the other large Aroids which have been tried in Battersea Park, the best are *Philodendron macrophyllum*, *P. Simsii*, and *Anthurium Hookeri*. Being rather tender they all require a very warm and well-sheltered position. W. Indies.

***Rhaponticum cynaroides.**—A hardy perennial from the Pyrenees, 3 ft. or more in height, with a rigid, simple, furrowed stem, and lobed or entire oblong radical leaves, covered underneath with silvery down. Flowers solitary, purple, in very large heads. It thrives in a deep, substantial, moist, but well-drained and free soil. This and, to a smaller extent, the following species are worthy of a place in full collections of hardy fine-leaved plants, for borders, the margins of groups, and, in the case of *R. cynaroides*, for isolation. Both are easily increased by division.

***Rhaponticum pulchrum.**—A hardy perennial from the Caucasus, with numerous simple stems 2 ft. or more in height, and much-divided, undulating, toothed leaves of an ashy or sea-green colour on the upper side, and whitish underneath. The flowers are borne in small solitary purplish heads late in summer. A very suitable subject for embellishing dry, arid, rocky positions. *R. scariosum* is another kind, useful for the same purposes as the preceding species.

*THE RHUBARBS.

The Rhubarbs, from their vigour and picturesqueness, are well worthy of cultivation among hardy, fine-leaved plants. They are so hardy that they may be planted in any soil, and afterwards left to take care of themselves. Their fine leaves and bold habit make them valuable ornaments for the margins of shrubberies (the best way is to plant one singly a few feet from the margin of the shrubbery, so that when they die down in autumn no blank may be seen), and for semi-wild places where a very free and luxuriant type of vegetation is desired. Though not particular as to soil, they enjoy it when it is deep and rich, and the more it is made so the better they will grow.

Rheum Emodi is undoubtedly the handsomest and most distinct of the genus in cultivation. The figure conveys an accurate idea of the outline of its leaves, and of its aspect when in flower. The large leaves have their veins red, which distinguishes it from any other species. It has a large and deep-feeding root, black on the outside, and yellow within. The flowers are very small, of a yellowish white. It comes up somewhat later than the common kinds, and is not by any means common, though it may be found in botanic gardens and nurseries where collections of herbaceous plants are formed. It may, like all the species, be increased by division, but a young plant should not be disturbed for several years after being planted. It is a native of Nepaul.

The palmated rhubarb, *Rheum palmatum*, is immediately distinguished from its cultivated fellows by its leaves

RHEUM EMODI.

Hardy herbaceous fine-foliaged Type.

being deeply cut into lobes. It is scarcely so ornamental or imposing as the fuller-leaved kinds, but is an interesting plant. I have seen it grown in some Irish gardens for culinary purposes, but do not remember to have noticed it in English kitchen-gardens. When well grown in deep and rather light and well-drained earth, the flowering-stem of this species attains a height of about nine feet. It is a native of Tartary, and well deserves a place. *Rheum Ribes* is a somewhat delicate species, a native of Southern Persia, which may be seen in one or two of our botanic gardens, and more frequently in those of France; but we fear it is not hardy enough to thrive well in these islands. The most remarkable known species is the Himalayan (*R. nobile*), which has its flower-stems beautifully clothed with large straw-coloured and pink-edged bracts, so as to form what may be termed a pyramid of leaves; but, as this species is not in cultivation, we need not describe it further. It would prove a very welcome addition to our collection of hardy plants.

The common rhubarbs are said by some of our authorities to have chiefly sprung from *Rheum Rhaponticum*, and some of them have also come from the wavy-leaved Rheum (*R. undulatum*). In any case, some of the garden varieties of rhubarb are worth planting for ornamental purposes. They have been so planted in Hyde Park, but in masses—not the proper way to employ them. Kinds deserving of notice are *R. australe*, *R. compactum*, *R. rugosum*, *R. hybridum*, Victoria rhubarb (a garden variety, with very large leaves and long red stalks), Myatt's Linnæus, and Prince Albert (also garden varieties, and splendid ornamental plants). Mr. Shirley

Hibberd says he has found Scott's Monarch to be the most imposing and ornamental of all the garden varieties.

***Rhus Cotinus** (*Venetian Sumach*).—A bushy shrub, about 6½ ft. high, with simple, smooth, shining green leaves, and a very remarkable feathery inflorescence of a deep red colour. It requires a dry, gravelly, warm soil, and will grow in the most arid positions. Where it thrives the effect of its peculiar inflorescence is very fine and distinct. It is used with most effect as an isolated specimen, though it would group very well with such plants as Lindley's Spiræa. S. Europe.

***Rhus glabra laciniata.**—This variety of the smooth or scarlet Sumach is a small shrub with compound leaves, growing from 4 ft. to 7 ft. high, a native of North America, with finely-cut and elegant leaves, the strongest being about a foot long when the plants have been established a year or two. When seen on an established plant, these leaves combine the beauty of those of the finest Grevillea with that of a fern frond, while the youngest and unfolding leaves remind one of the aspect of a finely-cut umbelliferous plant in spring. The variety observable in the shape, size, and aspect of the foliage makes the plant charming to look upon, while the midribs of the fully-grown leaves are red, and in autumn the whole glow off into bright colour after the fashion of American shrubs and trees. During the entire season it is presentable, and there is no fear of any vicissitude of weather injuring it. Its great merit is that, in addition to being so elegant in foliage, it has a very dwarf habit, and is thoroughly hardy. Plants three years old and undisturbed for the last two years are not more

RHUS GLABRA LACINIATA.

Hardy deciduous Shrub Section.

than eighteen inches high. The heads of some are branched, but these are not less elegant than when in a simple-stemmed state, so that here we have clearly a subject that will afford a charming fern-like effect in the full sun, and add graceful verdure and distinction to the flower-garden. When the flowers show after the plant is a few years old, they may be pinched off; but this need only be practised in the case of permanent groups or plantings of it. To produce the effect of a Grevillea or a fern on a small scale, we should of course keep this graceful Rhus small and propagate it like a bedding-plant. Like most other shrubs, it has a tendency to branch; but to fully enjoy the beauty of the leaves it is best to cut down the plants yearly, as then the leaves given off from the simple erect stem are much larger and more graceful. It will, however, be necessary to allow it to become established before treating it in this way, as it is at present comparatively new to our gardens. The figure, sketched early in August, represents a young plant little more than a foot high, which had been cut down to the ground during the spring of the past year, and proves that its full beauty may be enjoyed in a very small state. It may be most tastefully used in association with bedding-plants, or on banks in or near the rock-garden or hardy fernery, planting it in light sandy loam. The graceful mixtures and bouquet-like beds that might be made with the aid of such plants need not be suggested here, while of course an established plant, or groups of three, might well form the centre of a bed. Planting a very small bed or group separately in the flower-garden, and many other uses which cannot be enumerated here, will occur to those who have once tried it. Some hardy plants of fine foliage are either so rampant or so topheavy that they cannot be wisely associated with bedding-plants. This is, on the contrary, as tidy and tractable a grower as the most fastidious could desire. It would be a mistake to put such a pretty plant under or near rough trees and shrubs. Give it the full sun, and good free soil.

***Rhus vernicifera** is distinct from the preceding, and has fine leaves. It is a native of Japan, and the source of the best Japan varnish according to Thunberg. Useful for grouping with the preceding or other hardy shrubs of like character.

***Ricinus communis** (*Castor-oil Plant*). — When well grown in the open air, there is not in the whole range of cultivated plants a more imposing subject than this. It may have been seen nearly 12 ft. high in the London parks of late years, and with leaves nearly 1 yd. wide. It is true we require a bed of very rich deep earth under it to make it attain such dimensions and beauty; but in all parts, and with ordinary attention, it grows well. In warm countries, in which the plant is very widely cultivated, it becomes a small tree, but is much prettier in the state in which it is seen with us—*i.e.*, with an unbranched stem clothed from top to bottom with noble leaves. Soon after it betrays a tendency to develope side-shoots the cold autumn comes and puts an end to all further progress; and so much the better, because it is much handsomer in a simple-stemmed state than any other. The same is true of not a few other large-leaved plants—once they break into a number of side-shoots their leaf beauty is to a great extent lost. It is as easily raised from seed as the common bean, requiring, however, to be raised in heat. It should be sown about the middle of February, and the plants gradually hardened off so as to be fit to put out by the middle of May. The Ricinus is a grand plant for making bold and noble beds near those of the more brilliant flowers, and tends to vary the flower-garden finely. It is not well to associate it closely with bedding-plants, in consequence of the strong growth and shading power of the leaves, so to speak. A good plan is to make a compact group of the plant in the centre of some wide circular bed and surround it with a band of a dwarfer subject, say the Aralia or Caladium, and then finish with whatever arrangement of the flowering plants may be most admired. A bold and striking centre may be obtained, while the effect of the flowers is much enhanced, especially if the planting be nicely graduated and tastefully done. For such groups the varieties of the Castor-oil plant are not likely to be surpassed. East Indies.

The most notable varieties are *R. c. sanguineus*, the stem, leaf-stalks, young leaves, and fruit of which are of a blood-red colour ; *R. c. borboniensis*, which in southern climates often attains the extraordinary height of 26 ft. in one year ; *R. c. giganteus*, a very tall kind from the Philippine Islands.

Other kinds in cultivation are *R. Belot Desfougerès* (a very tall and branching kind), *R. viridis* (of a uniform lively green colour), *R. insignis*, *R. africanus*, *R. africanus albidus*, *R. minor*, *R. hybridus*, *R. microcarpus*.

The better and richer the soil, and the warmer the position, the more vigorous will be the growth of any of the above. Copious watering in summer is indispensable.

***Rumex Hydrolapathum.**—A very large native water-plant of a size and habit sufficiently striking to entitle it to a place amongst ornamental subjects by the water-side. The radical long-stalked leaves, which are sometimes 2 ft. or more in length, form erect tufts of a very imposing character. The flowering-stem is frequently 6 ft. in height, and bears a very large, dense, pyramidal panicle of a reddish or olive-fawn colour. The plant is most effective in autumn, when the leaves change to a lurid red colour, which they retain for some time.

***Saccharum ægyptiacum.**—A vigorous perennial grass, forming ample tufts of reed-like downy stems 6½ ft. to 13 ft. high, and clothed with very graceful foliage, well adapted for ornamenting the margins of pieces of water, the slopes and other parts of pleasure-grounds, etc., in a warm position. In our climate it does not flower, but even without its fine feathery plumes it is a pretty plant from its foliage and habit alone. Easily and quickly multiplied by division in spring; the offsets to be started in a frame or pit. When established they may be planted out in May or June. N. Africa.

***Sagittaria sagittifolia.**—A British water-plant, affording the most remarkable example of the arrow-shaped leaf to be met with among hardy plants. These leaves stand erect, from 1 ft. to 1½ ft. above the water, and from the middle of the tuft the flowering-stem rises in August to the height of 1½ ft. to 2½ ft. The flowers are of a pale rosy-white colour. There is a variety with double flowers (*S. sagittifolia flore pleno*), which resemble the flowers of the double Rocket. Both the double and single kinds should have a place among water or bog plants.

SEAEORTHIA ELEGANS (10 ft.).

Conservatory Palm; standing well in the open air in summer.

***Salvia argentea** (*S. patula*).—A handsome biennial about 3½ ft. high, forming broad spreading rosettes of large, oval, heart-shaped leaves, densely covered with long silky hairs of a silvery whiteness, and bearing large panicles of white flowers. An excellent subject for grouping on grass-plats or the uneven parts of pleasure-grounds. Where the effect of the foliage only is desired the flower-stems should be pinched off as soon as they appear; the leaves will then preserve their freshness and silvery colour throughout the year. A light, sandy, or gravelly soil is the most suitable. Easily propagated by sowing in autumn, and keeping the seedlings in a cold frame through the winter, giving them air as often as possible, and watering very moderately. This plant is most effective during its first summer, and before it makes an attempt to flower. S. Europe.

***Scirpus lacustris** (*Bulrush*).—This giant rush sends up numerous smooth green stems as thick as the finger at the base, and from 3 ft. to 8 ft. high. In still water the bases of these are covered with leafless sheaths, but in running water the uppermost sheath produces at its extremity a leaf of several inches in length, in addition to which numerous barren tufts of leaves, often of great length, and resembling those of *Sparganium affine*, are also produced. When in flower, the stems are crowned with short, umbel-like, chestnut-coloured panicles. It is very effective on the margins of lakes or streams, associated with other tall and imposing aquatic plants.

Seaforthia elegans.—One of the most beautiful of the Palm family, from the northern parts of Australia, where it attains a height of about 30 ft., but in this country seldom arrives at more than half its full size. The leaves are from 2 ft. to 10 ft. in length, and are divided into numerous narrow leaflets from 1 ft. to 1½ ft. long, and of a dark green colour. The whole plant is perfectly smooth, and is one of the finest subjects in cultivation for the conservatory, greenhouse, or subtropical garden. It may be placed in the open air from the middle or end of May until the beginning of October. It is too scarce as yet to be procurable by horticulturists generally, but should be looked for by all who take an interest in these matters and have a house in which to grow it. It stands well in the conservatory during the winter, though generally kept in the stove, where of course it grows beautifully. There are hardier kinds—the dwarf Fan-palm for example—but on the whole none of them are so valuable as this.

Senecio Ghiesbreghtii.—A handsome Mexican plant with stout, round, spotted stems, 3 to 4 ft. high, and large oval-oblong, thick, coarsely-toothed leaves of a light green colour, and slightly rolled down at the margin. Flowers small, yellow, very numerous, in corymbose clusters of enormous size. A useful plant for isolation on grass-plats, or for beds, etc. Young plants are to be preferred for this purpose, as the old ones are apt to become bare and ragged-looking at the base. Plant in a mixture of peat soil and free loam at the end of May. Multiplied by cuttings in winter, struck under glass in a temperate heat in early spring.

Senecio Petasites (*Cineraria platanifolia*).—Another Mexican species, nearly 3½ ft. high, with a stout, half-shrubby, slightly-branching stem, and large, dark-green, roughish, lobed leaves. It requires a substantial, but free and cool soil, and may be multiplied at almost any season by cuttings. It requires greenhouse protection in winter, grows freely in the open air planted out in early summer, and is suitable for beds or groups associated with the medium-sized subjects.

***Seseli gummiferum.** (*Silvery Seseli*).—An umbelliferous plant with elegantly-divided leaves of a peculiarly pleasing glaucous or almost silvery tone. I am not sure whether this plant is perennial or not, and it is not hardy on cold soils, having perished during several of our most severe winters, but it is so unique in its way that some persons might like to grow it, and if so the best position is on dry and sunny banks, or raised beds or borders. It is one of the few subjects that are at once fern-like and silvery, and if plentiful enough might be used to form charming edgings. It is not difficult to raise from seed, which should be sown soon after being gathered. As it is liable to perish in winter, perhaps the best way to deal with it would be to put it out annually at the end of spring, raising it and keeping it in frames for this purpose; and to secure seeds a few plants might be left in 10 in. or 12 in. pots, so that the seed might be gathered under cover. Doubtless, however, continental raisers can supply it. S. Russia.

***Silphium integrifolium.**—This composite has a rough, vigorous, rigid, slightly four-angled, grooved stem, from 2 ft. to 4 ft. high, and leaves from 3 ins. to 5 ins. long, all opposite, lanceolate-ovate, entire, tapering to a sharp point from a roundish heart-shaped and partly-clasping base, and covered with a roughish down. The flower-heads are of a greenish-yellow, on short stalks, in a close, forking corymb. A variety (*S. læve*) has the leaves and stem smooth, or nearly so. This and the following kinds are only suited for the rougher parts of the pleasure-ground, and by wood-walks, etc.; they will be seen to greatest advantage in rich and deep soil, but will grow in any kind. Division. N. America.

***Silphium laciniatum** (*Compass-plant*).—A vigorous perennial with a stout, round stem, often upwards of 8 ft. in height. The leaves, which are collected chiefly about the base of the plant, are large, wrinkled, and deeply-divided into lance-shaped, pointed segments, and fringed with white hairs. The stem-leaves are few, much smaller, and opposite. Flowers of a fine yellow with a brownish centre, in large, few, solitary, short-stalked, horizontal or drooping heads, which have the peculiarity of facing to the east. Division. N. America.

***Silphium perfoliatum** (*Cup-plant*).—A robust-growing North American perennial from 4 ft. to 8 ft. in height, with a square stem and broad, opposite, oval, lance-shaped, coarsely toothed leaves, 6 ins. to 15 ins. long, rough on both sides, the upper ones united at their bases; the lower ones abruptly narrowed into winged leaf-stalks, which are also united at their bases. Flower-heads about 2 ins. across, with a greenish-yellow disk and a yellow ray. Does best in a deep, free, well-drained, sandy soil, but will grow well when the ground is slightly moist, especially in warm, but not too shady, positions. Division.

***Silphium terebinthinaceum** (*Prairie-Dock*).—A large kind with smooth slender stems from 4 ft. to 10 ft. high, panicled at the summit, and bearing many small heads of light yellow flowers. The leaves are ovate-oblong, thick and rough, especially beneath, and from 1 ft. to 2 ft. long, on slender stalks. A variety (*pinnatifidum*) has the leaves deeply cut or pinnatifid. This species is remarkable for its strong turpentine odour. Division. N. America.

***Silphium trifoliatum.**—This species has a smooth, often glaucous, rather slender stem, from 4 ft. to 6 ft. high, branching above. Leaves lance-shaped, pointed, entire or scarcely serrate, short-stalked, in whorls of three or four, the uppermost ones opposite. Flower-heads yellow, more than 2 ins. across, on long stalks, and forming loose panicles. Division. N. America.

***Silybum marianum** (*Milk-thistle*).—A very robust and vigorous-growing native biennial, 5 ft. or more in height, of strikingly handsome appearance, and well deserving to be associated with other large fine-foliaged plants. Its leaves are of very great size, variously cut and undulated, tipped and margined with scattered spines, and of a bright glistening green colour marbled and variegated with broad white veins. Easily raised from seed, and thrives in almost any kind of well-drained soil. Additional vigour and development may be thrown into the foliage by pinching off the flower-stems on their first appearance. If a few plants are raised in the garden and planted out in rough and somewhat bare places or banks, etc., this will soon establish itself permanently.

Silybum eburneum is a more tender species, very closely resembling the above, but with spines which appear as if made of ivory. It is also more constantly biennial, and in consequence its leaves are almost always in the rosette stage throughout the first year. It is somewhat tenderer than *S. marianum.* Algeria.

The Solanums.—This family, so wonderfully varied, affords numerous species that look graceful and imposing in leaf when in a young and free-growing state. In selecting examples from this great genus we must be careful, as our climate is a shade too cold for some of the kinds grown on the continent, and many of them are of too ragged an aspect to be tolerated in a tasteful garden. Half a dozen species or so are indispensable, but there is quite a crowd of narrow-leaved and ignoble ones which may well be passed over.

Most of these plants may be raised from seed, while they are also freely grown from cuttings, which struck in February will make good plants by May. All the kinds named are suitable for association with the larger-leaved plants, though they do not as a rule attain such height and vigorous development as those of the first rank, like the Ricinus. As a rule, temperate-house treatment in winter is required, and they should be planted out about the middle or end of May, in rich light soil, a warm position, and perfect shelter. *S. marginatum*, planted in a very dwarf and young state, furnishes a most distinct and charming effect: it should be planted rather thinly, so that the leaves of one plant may not brush against those of another. If some very dwarf plants are used as a groundwork, so much the better; but the downy and silvery leaves of this plant are sure to please without this aid. It is very much better when thus grown than when permitted to assume the bush form.

Solanum betaceum.—A small tree from South

SOLANUM ROBUSTUM.

Tender Section; making vigorous growth during the summer months.

America, which in our climate attains a height of nearly 10 ft. if taken up in autumn and kept through the winter in a house. The stems are stout, smooth, and fleshy. The leaves, which resemble those of the Beet, are of an oval, pointed shape, and of a deep green colour, tinged with violet in the variety *purpureum.* The flowers are small, rose-colour, in pendent cyme-like clusters, and are succeeded by fruit of the shape and size of a fowl's egg, which become of a fine deep scarlet colour during the winter. Some varieties have flowers tinged with purple and fruit striped with brown. May be placed to great advantage in groups in round beds with dwarfer plants or shrubs at the base, or with climbing plants ascending the stems, but is much better isolated on slopes, etc. It is a vigorous grower, and should have rich soil.

Solanum crinitipes.—A slow-growing woody species with undivided oval leaves somewhat more than a foot long: the young stems and flower-stalks being densely covered with chaffy hairs somewhat like those of a fern. This I have not seen thrive so well in England as the preceding kind, but it is well worthy of trial in full collections in the southern counties. S. America.

Solanum crinitum.—A vigorous-growing species from Guiana, 5 ft. or more in height, with stout stems, set with short strong spines and dense long hairs. It has very large, soft, hairy, spreading, roundish leaves, which in good soil attain a length of 2½ ft.: the upper surface of a tender green colour with violet veinings set with spines, as are also the leaf-stalks; the under side whitish and more thickly furnished with spines. The hairs and bark on the upper portion of each petiole are of a purplish hue, and, on the lower part, of a light pale green, by which the plant may be readily recognised. The flowers are very large and white. Berries roundish, villose, and twice or thrice as large as a cherry. This I have seen attain a very remarkable development in sheltered warm spots in the south of England. It is fine in medium-sized groups.

Solanum hyporhodium.—A fine branching kind from Venezuela, with a stout stem about 5 ft. high, and branches armed with short thinly-scattered spines. The leaves, which attain a length of nearly 2½ ft., are oval, with angular sinuated lobes, the upper surface being of a fine green colour with white veins, and the under side of a violet-red and downy. When young the hue of the leaves is exceedingly lively. The flowers are borne in almost lateral cymes and are of a rosy-white colour with yellow stamens. This plant is sometimes sold as *S. discolor* and *S. purpureum*, but is quite distinct from them.

Solanum Karstenii.—This, which is more commonly known as *S. callicarpum*, is a robust, slightly branching, arborescent shrub about 5 ft. high, covered with long hairs interspersed with spines and of a general variable greyish-violet hue. The leaves are oval, broad, angular, heart-shaped at the base, and 2 ft. or more in length. The flowers are large, of a fine delicate violet colour, and borne in crowded, almost one-sided clusters. This plant is best isolated, as when placed in close groups the leaves of the associated subjects are apt to tear it. Venezuela.

Solanum lanceolatum.—This is the best kind for blooming qualities. The foliage, which is somewhat fluffy and willow-like, possesses no marked character, but the mauve-coloured flowers are borne abundantly in clusters, each containing 20 or more blooms : the stamens, being of an orange colour, add to the effect. There are a dozen or more species that flower freely but have little beauty of leaf: among the best of these is *S. Rantonnettii*, which has very pretty dark-purple flowers, more than an inch across, with an orange centre. It forms a neat bush, and flowers freely in the southern counties, in warm sunny spots and on light soils. Mexico.

Solanum macranthum.—A fine species from Brazil, confessedly one of the best kinds in cultivation, and somewhat resembling *Polymnia grandis.* It grows nearly 7 ft. high in one year, with a stout, simple, spiny stem of a deep shining green with grayish spots, and sparsely armed with very strong shortish spines. The leaves are elegant and deeply cut, some of them over 2½ ft. long, falling gracefully earthwards, of a light green on the upper surface, with red veinings, the under side having a reddish hue. The flowers, seldom seen with us, are of a fine violet colour, and grow in corymbs. It will not attain its full character and large dimensions in cold places, and should therefore have as warm positions as possible. Increased by cuttings struck in February: they are fit to plant out in May.

Solanum marginatum.—A vigorous-growing, erect, branching and bushy species from Abyssinia, 3 ft., or more, in height. The leaves are somewhat oval, with a bluntly sinuated margin; the upper surface smooth, of a brilliant green with a white silvery border, and the under side covered with a white satiny down. The flowers are white, with orange stamens, pendulous, very numerous, in clusters. For the positions suited to this plant see the introductory remarks on the genus.

Solanum Quitoense.—A half-shrubby native of Peru and the neighbourhood of Quito, seldom growing higher than 3½ ft. in cultivation. The stem is spineless, covered with a soft down, and of a delicate green colour suffused with violet, which exhibits iridescent changes. The leaves are broad, stalked, obcordate, with toothed angles, and of a fine green colour, with violet downy veins. Flowers rather large, white, tinged with lilac on the top when in bud, in short clusters. It requires a warm position and a warm season to bring out its best qualities.

Solanum robustum.—A Brazilian species with a vigorous much-branching stem more than 3 ft. high, and furnished with very sharp and strong spines and densely-set, long, reddish, viscous hairs. The leaves, which are very large, are of a rich brown colour on the upper surface and oval-elliptical in form, with 8 or 9 oval-acute lobes, the upper ones nearly triangular; and the midrib and principal veins, which are of a brown colour, are closely set with spines similar to those on the stem. The flowers are white, with orange stamens, and are borne in unilateral clusters. The berries are round, of a brown colour, and the size of a small cherry. As a foliage-plant this is a subject of considerable merit, and one of those most suitable for our climate. It requires a warm sunny aspect in a position which will be at the same time airy and sheltered from strong winds.

Solanum Sieglingii.—A large and handsome kind, which forms a small tree about 13 ft. high after some

SOLANUM WARSCEWICZII.

Tender Section; making noble leaves in the open garden in summer.

years' growth. The foliage is of a light-green colour, tinged here and there with rose, and sparsely armed with spines; the young unfolded leaves are slightly tinged with violet. Flowers numerous, small and white, appearing when the plant is two or three years old. A good kind which has been little tried in England. Venezuela.

Solanum Warscewiczii.—A very fine and ornamental kind, resembling *S. macranthum*, but with a lower and more thickset habit, and branching more at the base. The leaf-stalks also, and upper branches, are of a red colour, glandular, and scaly; and the flowers are white and small. The stem is armed with strong slightly recurved spines, and both the stems and the petioles of the leaves are covered with a very dense crop of short stiff brown hairs scarcely rising above the skin. This is one of the handsomest and best kinds we have.

Sonchus laciniatus.—A very graceful composite plant, from Madeira, with a stout stem, growing to a height of more than 5 ft., and large deeply-cut leaves with linear-lance-shaped segments. Flower-heads yellow. When grouped on grass-plats, or open spaces in pleasure-grounds, the fine foliage of this plant is seen to very great advantage; but being so slender and delicate the plants must be placed where they may be seen. It should be planted out at the end of May, and thrives best in rich, substantial soil, in a warm sunny position. Very numerous varieties, with the leaves variously divided and of various shades of green, have been advertised in catalogues under specific names, as *S. lyratus*, *S. gummiferus*, etc., etc. Many of these are quite as charming as the type, and are well adapted for the same uses.

***Sorghum halepense.**—A handsome hardy grass from S. Europe, N. Africa, and Syria, with an erect stem about 3½ ft. high, and broad flat leaves more than 1 ft. long, chiefly collected round the base of the plant. It is most attractive when in flower in the end of summer, the inflorescence consisting of a dense panicle of purplish awned flowers. Suitable for isolation, groups, or borders.

Sparmannia africana.—A beautiful flowering stove-shrub from 3 ft. to 12 ft. high, very much resembling a Malva in habit, with long-stalked, heart-shaped, lobed leaves, clothed with soft down, and numerous pretty white flowers produced in stalked umbels. It thrives freely in the open air in the south of England, from the end of May to October, if planted in rich light soil and in warm positions. Cape of Good Hope.

***Spiræa Aruncus.**—This is a remarkably handsome and effective plant, from 3½ ft. to 5 ft. high, with elegantly-divided leaves, which bear some resemblance to the fronds of certain ferns. The flowers are white, and are disposed above the foliage in graceful, airy plumes. A cool, peaty soil, and a slightly-shaded position, are best suited for this plant, and it may be placed with advantage on slopes with a north aspect, the banks of streams or pieces of water, in glades, and thinly-planted shrubberies, etc. Division. Siberia.

***Spiræa Filipendula.**—A hardy, native perennial, with elegant foliage and handsome flowers. The leaves are mostly radical, very finely cut, and form a loosely-spreading rosette. The flower-stems rise to a height of 1½ ft. to 2 ft., and are terminated by dense panicles of rosy-white flowers. There is a fine variety with double flowers. This plant is included here only in consequence of the resemblance of its leaves to a pinnate-leaved fern. By pinching off the flowers it may be used with good effect as a green, fern-like edging plant, and it is pretty in borders. Division in winter or spring.

***Spiræa (Hoteia) japonica.**—A handsome, herbaceous perennial, forming rich tufts of dark shining green much-divided leaves, which have a somewhat fern-like appearance. These tufts are usually from a foot to 16 ins. high. The flowers are very freely produced in graceful panicles, of which the bracts, little flower-stems, and all the ramifications are, like the flowers, white. It is particularly fond of a sandy peat, or very sandy loam, a sheltered position, and moist soil. Multiplied by division of the tufts in spring or the end of summer. Japan.

***Spiræa Lindleyana.**—A graceful shrub, with erect stems, from 6½ ft. to nearly 10 ft. high, and large compound leaves, with finely-toothed leaflets. Flowers late in summer, white, in very large and handsome terminal panicles. This well-known plant is second to none for its grace and distinctness, both of foliage and flower. It is a native of the Himalayas, and easily procured in our nurseries; it should receive far more attention than the majority of our shrubs do, and should be employed both in a young and fully-grown state in and near the flower-garden. Few things, tender or hardy, known in our gardens, afford a better effect than may be obtained from this.

It is probably one of those plants which would look exceedingly effective if trained to a single stem and cut down every year, as recommended for the Ailantus and the Paulownia; but I have had no experience of it in this way, and its natural habit is sufficiently graceful.

Stadmannia Jonghei.—A tall and stately foliage-plant from Australia, where it attains the dimensions of a small tree, with dark shining green pinnate leaves; the divisions oblong-pointed, with serrated margins, and of a paler colour underneath. Bears the open air of the southern counties in summer well, if placed in sunny and sheltered spots.

***Statice latifolia.**—A hardy and very ornamental herbaceous perennial from Russia, with broad leaves, which form a rosette or tuft more or less spreading. The flower-stem is more than 2 ft. high, and very much branched; the branches commencing at from 4 ins. to 8 ins. above the ground, and forming a large and exceedingly handsome panicle of flowers of a light-blue colour, tinged with the greyish hue of the numerous membranous bracts and thin dry calyces. A well-drained, sandy soil, in an open sunny position, is the best for this plant, which, however, grows in any ordinary garden-soil, and is admirably adapted for naturalisation or grouping with the acanthuses, tritomas, etc., the effect of the inflorescence being very remarkable.

***Stipa pennata** (*Feather-grass*).—This plant, which at other times is hardly to be distinguished from a strong, stiff tuft of common grass, presents, in May and June, a very different appearance, the tuft being then surmounted by numerous flower-stems, nearly 2 ft. high, gracefully arching, and densely covered, for a considerable part of their upper extremity, with long, twisted, feathery awns. It loves a deep, sandy loam, and may be used with fair effect in groups of small plants, or isolated; but its flowers continue too short a time in bloom to make it very valuable away from borders.

***Struthiopteris germanica.**—One of the most elegant hardy ferns, with fronds resembling ostrich-plumes in shape, nearly 3 ft. long, and arranged in a somewhat erect, vase-like rosette. It is particularly suited for the embellishment of the slopes of pleasure-grounds, cascades, grottoes, and rough rockwork, the margins of streams and pieces of water, and will thrive in moist and deep sandy soil, either in the full sunshine or in the shade. *S. pennsylvanica* very closely resembles *S. germanica*, the chief point of difference being the narrowness of the fertile fronds of the former species. Both kinds will prove very effective in adding beauty of form to a garden, and should by no means be confined to the fernery proper. Central Europe.

***Tamarix.**—These very elegant hardy shrubs may be used with excellent effect in the flower-garden and pleasure-ground, though they are at present seldom employed in these places. *T. gallica* or *anglica* is found apparently wild in several parts of the south of England, and other kinds, such as *germanica*, *parviflora*, *tetrandra*, *spectabilis*, and *indica*, are also in cultivation. In the neighbourhood of Paris *T. indica* thrives very freely, and forms beautiful hedges, but is cut down by frost during some winters. It would probably do better in the south of England. The plants have minute leaves and very elegantly-panicled branches, which gives them a feathery effect, somewhat like that of the most graceful conifers, and, if possible, more elegant: the roseate panicles of small flowers are also very pretty. A finer effect would be obtained from these shrubs by isolating them on the grass than in any other way.

***Tanacetum vulgare var. crispum.**—A very elegant variety of the common tansy, much dwarfer in stature, and with smaller emerald-green leaves, which are very elegantly cut, and have a crisped or frizzled appearance. It is quite hardy, and forms an effective ornament on the margins of shrubberies, near rockwork, etc. It does best fully exposed, and probably the only way in which it can be benefited after planting—in deep and rather moist soil it does best, but will grow "anywhere"—is by thinning out the shoots in spring, so that each remaining one shall have free room to suspend its exquisite leaves; thinned thus, it looks much better than when the stems are crowded, and of course, if it is done in time, they individually attain more strength and dignity. The flowers should be pinched off before they open. Britain.

Thalia dealbata.—This is one of the finest aquatic plants which we can employ in the embellishment of pieces of water, streams, etc. In a warm and sheltered position, and on a substantial and rich bottom, it grows vigorously, sometimes attaining a height of 6 ft. The best mode of growing it is in pots or tubs pierced with holes, in a mixture of stiff peat and clayey soil, with a portion of river-mud and sand. In winter these pots or tubs may be submerged to a greater depth, and the plants be thus effectually protected. It would not attain the above size out of doors except in warm places in the southern counties, in which it might be planted out directly without taking the precautions above described.

It is generally grown in the stove in this country. N. America.

***Thalictrum minus.**—One of the most elegant-leaved of our native plants, forming compact, roundish bushes, from a foot to 18 ins. high, very symmetrical, and of a slightly glaucous hue. It may be grown in any soil, and requires only one little attention, namely, to pinch off the slender flower-stems that appear in May and June. Not alone in its aspect, as a little bushy tuft, does it resemble the "Maidenhair Fern," as *Adiantum cuneatum* is often called, but the leaves are almost pretty enough to pass, when mingled with flowers, for those of the fern; they are also stiffer and more lasting than fern-leaves, and are well suited for mingling with vases of flowers, etc. There are probably several "forms" or varieties of this plant. It would look very pretty isolated in large tufts as an edging, or in borders, or in groups of dwarf subjects. Easily increased by division.

***The Tritomas.**—So hardy, so magnificent in colouring, and so fine in form are these plants, that we can no more dispense with their use in the garden where beauty of form as well as colour is to prevail, than we can with the noble Pampas grass. They are more conspicuously beautiful, when other things begin to succumb before the gusts and heavy rains of autumn, than any plants which flower in the bright days of midsummer. It is not alone as component parts of large ribbon-borders and in such positions that these grand plants are useful, but in almost any part of the garden. Springing up as a bold, close group on the green turf, and away from brilliant surroundings, they are more effective than when associated with bedding plants; and of course many such spots may be found for them near the margins of the shrubberies in most pleasure-grounds. It is in an isolated group, flaming up amid the verdure of trees and shrubs and grass, that their dignified aspect and brilliant colour are seen to best advantage. However, tastefully disposed in the flower-garden, they will prove generally useful, and particularly for association with the finer autumn-flowering herbaceous plants. A most satisfactory result may be produced by associating the Tritomas with the Pampas grass and the two Arundos, the large *Statice latifolia*, and the strong and beautiful autumn-flowering *Anemone japonica alba*, which is peculiarly suited for association with hardy herbaceous plants of fine habit, and should be in every garden where a hardy flower is valued.

The Tritomas are not fastidious as to soil, and with a little preparation of the ground may be grown almost anywhere. They thrive with extraordinary vigour and freedom where the soil is very sandy as well as rich and deep, and are readily multiplied by division.

As every garden should be embellished by well-developed specimens or groups of these fine plants, those who have very poor and thin, or pure clay soils, would do well to excavate the ground to the depth of 2 ft. or 3 ft., and fill in with good rich loam. When the soil is deep, no watering will be required.

***Tritoma Burchelli.**—This kind is distinguished by the lighter green of its leaves, by its black-spotted flower-stem, and especially by the colour of its flowers, which are crimson at the base, passing into carmine in the middle, and pale-yellow or greenish at the tips. There is a variety which has the leaves variegated or striped with white, but it is somewhat tender and rare.

***Tritoma glauca.**—A dwarfer kind than *T. Uvaria*, with leaves of a sea-green colour, and very large spikes of scarlet-and-yellow flowers, which, when in bud, are hidden by long, sea-green bracts, streaked and rayed with white. There is a scarce variety with recurved leaves (*T. g. recurvata*), which has somewhat of the habit of a Bromelia. S. Africa.

***Tritoma præcox.**—A recently-introduced, handsome, hardy perennial, with very much the habit of *T. Uvaria.* The flower-stem grows from 20 ins. to 2 ft. high, and the flowers, which are produced about the middle of May, are of a bright-red colour when exposed to the full sun, and of a bright-yellow when grown in the shade. The leaves are fully 2 ft. long, sharply keeled, and with toothed edges. S. Africa.

***Tritoma Uvaria.**—A very ornamental and well-known kind from S. Africa, forming thick tufts of linear, erect leaves. It is a vigorous grower, and small specimens have been known in three years to form tufts from 3 ft. to 4 ft. through, bearing from 50 to 100 flower-spikes. The flowering-stems are about 3½ ft. in height, and the flowers are borne in dense conical clusters at the top. The upper part of the cluster, containing the young flowers, is of a coral-red colour, the lower part yellow, all the flowers gradually changing to this colour. Other varieties in cultivation are—*T. U. grandis* or *grandiflora*, which is much taller than the preceding kind, with stouter stems and larger flower-spikes; *T. U. Rooperi*, which only differs from the type in being somewhat dwarfer in habit and having softish or flaccid leaves, frequently falling forward; it also flowers later; and *T. U. Lindleyana*, which has erect, very rigid leaves, and more deeply-coloured flowers than the type.

Tupidanthus calyptratus. — A noble subtropical plant from Bengal, standing in the open air from the beginning of June till October without the slightest injury. The leaves are large, deeply-divided, and of a dark shining green colour. It requires stove treatment in winter and spring, and is suitable for beds or planting singly.

***Typha latifolia** (*Reed-Mace*). — A native aquatic plant, growing in tufts of 2-rowed flat leaves from 1½ ft. to 2 ft. long, and 1 in. or 1½ in. wide. From the centre of each tuft springs a stem 6 ft. or 7 ft. high, which in the flowering season is terminated by a close cylindrical spike 9 ins. long, and of a dark-olive colour, changing to a brownish-black as it ripens. This is one of the most striking and ornamental of our British water-plants, and may be used with excellent effect grouped with such subjects as the Great Water-Dock.

***Typha angustifolia** resembles the preceding species in all respects except in the size of its leaves and spike. The leaves are about ½ in. wide and the spike about ½ in. in diameter, and something shorter than that of *T. latifolia.* Of the two it is perhaps the more graceful in aspect.

Uhdea bipinnatifida. — This is one of the most useful plants in its class, producing a rich mass of handsome leaves, with somewhat the aspect of those of the great cow-parsnips, but of a more refined type. The foliage has a slightly silvery tone, and the plant continues to grow fresh and vigorously till late in autumn.

It is well suited for forming rich masses of foliage, not so tall, however, as those formed by such things as Ricinus or Ferdinanda. It is freely propagated by cuttings taken from old plants kept in a cool stove, greenhouse, or pit during the winter months, and placed in heat to

Uhdea bipinnatifida.

afford cuttings freely in early spring. Under ordinary cutting treatment on hotbeds or in a moist warm propagating house, it grows as freely as could be desired, and may be planted out at the end of May or the beginning of June. Mexico.

Uhdea pyramidata.—This kind has been less cultivated in England than the preceding, from which it is distinct in appearance. It is of a lighter and fresher green, and inclined to grow larger in habit, having more of the aspect of a Malva in foliage. Useful for the same purposes as the preceding kind, but not so valuable.

***Veratrum album** (*White Hellebore*).—A handsome, erect perennial of pyramidal habit, 3½ ft. to 5 ft. high, with curiously plaited leaves 1 ft. long and 6 ins. to 8 ins. broad, regularly alternating on the stem and overlapping each other at the base. The flowers, of a yellowish-white colour, are borne in numerous dense spikes on the top of the stem, forming a large panicle. The leaves being handsome, it is worth a place in full collections of fine-foliaged hardy herbaceous plants, and would look to best advantage in small groups in the rougher parts of the pleasure-ground and by wood-walks. Thrives best in peaty soil, and is best multiplied by division, as the seed is very slow and capricious in germinating, sometimes not starting until the second year, and it is some years before the seedlings are strong enough to flower. The root of this plant is exceedingly poisonous. *V. nigrum* differs from *V. album*, in having more slender stems, narrower leaves, and blackish-purple flowers. *V. viridiflorum* resembles *V. album* in every respect, except that its flowers are of a lively green colour. France.

***Verbascum Chaixii.**—Most of us know how very distinct and imposing are the larger Verbascums, and those who have attempted their culture must soon have found out what far-seeding things they are. Of a biennial character, their culture is most unsatisfactory: they either migrate into the adjoining shrubbery or disappear altogether. The possession of a fine perennial species must therefore be a desideratum, and such a plant will be found in *Verbascum Chaixii*. This is fine in leaf and stature, and produces abundance of flowers. The lower leaves grow 18 ins. or 20 ins. long, and the plant when in flower reaches a height of 7 ft. or 8 ft., or even more when in good soil. It is a truly distinct subject, and may, it is to be hoped, ere long be found common in our gardens and nurseries. Like the preceding, but grown under the name *V. vernale*, is a kind I saw in the Jardin des Plantes at Paris, and introduced into cultivation in England; but it is as yet scarce.

Verbesina gigantea.—An ornamental shrub from Jamaica, about 6½ ft. high, forming, when young, a very pleasing subject for decorative purposes, its round green stems being covered with large, winged, pinnate leaves of a glistening delicate-green colour, and very elegant outline. Suitable for rich beds or groups; and should be planted out at the end of May or early in June. *V. pinnatifida* is a rough, half-shrubby species with a winged stem and woolly oval leaves with lobed or toothed margins; they are larger than those of the preceding species, growing 3 ft. long by 14 ins. broad in the first year. Both species require hothouse treatment in winter, and are multiplied by cuttings in early spring. Young plants are to be preferred for effect, and will be much the better for as warm and sheltered a position and as rich and light a soil as can be conveniently given them.

Wigandia macrophylla (*caracasana*).—This noble

plant, a native of the mountainous regions of New Granada, is, from the nobility of its port and the magnificence of its leaves, entitled to hold a place among the finest plants of our gardens. Under the climate of London it has made leaves which have surprised all beholders, as well by their size as by their strong and remarkable veining and texture. It will be found to succeed very well in the midland and southern counties of England, though too much care cannot be taken to secure for it a warm sheltered position, free good soil, and perfect drainage. It may be used with superb effect either in a mass or as a single plant. It is frequently propagated by cuttings of the roots, and grown in a moist and genial temperature through the spring months, keeping it near the light so as to preserve it in a dwarf and well-clothed condition; and, like all the other plants in this class, it should be very carefully hardened off previous to planting out at the end of May. It is, however, much better raised from cuttings of the shoots, if these are to be had. It may also be raised from seed. *W. macrophylla* has the stems covered with short stinging hairs, and bearing brownish viscid drops, which adhere to the hand like oil when the stem is touched.

W. Vigieri is another fine kind of quick and vigorous growth, and remarkable habit. In the beginning of September, 1867, I measured a specimen with leaves 3 ft. 9 ins. long, including the leaf-stalk, and 22 ins. across; the stem, nearly 7 ft. high and 3 ins. in diameter, bearing a column of such leaves. It is known at a glance from the popular and older *W. macrophylla*, by the leaves and the stems being covered in a much greater degree with glossy,

WIGANDIA MACROPHYLLA. (*W. caracasana*).

Tender Section; making noble leaves in the open air in summer.

slender, stinging bodies. These are so thickly produced as to give the stems a glistening appearance. *W. urens* is another species often planted, but decidedly inferior to either of the foregoing, except in power of stinging, in which way it is not likely to be surpassed.

Woodwardia. — This noble genus of ferns is of great and peculiar use in the subtropical garden, where their broad and beautifully arching fronds make very effective objects, especially when planted in a vase, on the top of a stump, or small mound; a little above the level of the eye. The principal species are: *W. orientalis*, *W. radicans*, *W. japonica*, *W. virginica*, and *W. areolata*. Of these *W. orientalis* and *W. areolata* are hardy, and the others nearly so. They may be used as effectively in the conservatory in winter as in the open garden in summer.

Xanthoso sagittæfolium. — A Brazilian plant with very much the habit and appearance of *Caladium esculentum*, but not so valuable, having arrow-shaped leaves, of a dark-green colour, supported on rather slender stalks. Another equally handsome and large species is *X. violaceum*, the leaves and leaf-stalks of which are suffused with a delicate violet hue, slightly inclining to hoariness. Positions and treatment similar to those recommended for *Caladium esculentum.* They should only be tried in the warmer parts of the country, and not be placed in the open air till the beginning of June.

Yuccas.—Among all the hardy plants ever introduced into this country, none surpass for our present purpose the various kinds of Yucca, or "Adam's Needle," as it is commonly called. There are several species hardy and well suited for flower-garden purposes, and, more advantageous still, distinct from each other. The effect afforded by them, when well developed, is equal to that of any hothouse plant that we can venture in the open air for the summer, while they are green and ornamental at all seasons. They may be used in any style of garden, may be grouped together on rustic mounds, or in any other way the taste of the planter may direct. If we had but this family alone, our efforts to produce an agreeable effect with hardy plants could not be fruitless. The free-flowering kinds, *filamentosa* and *flaccida*, may be associated with any of our nobler autumn flowering plants, from the Gladiolus to the great *Statice latifolia.* The species that do not flower so often, like *pendula* and *gloriosa*, are simply magnificent as regards their effect when grown in the full sun and planted in good soil; and I need not say bold and handsome groups may be formed by devoting isolated beds to Yuccas alone. They are mostly easy to increase by division of the stem and rhizome; and should in all cases be planted well and singly, beginning with healthy young plants, so as to secure perfectly developed specimens.

Yucca aloïfolia.—A fine and distinct species, with a stem when fully developed as thick as a man's arm, and rising to a height of from 6 ft. to 18 ft. Leaves numerous, rigidly ascending, dark-green, with a slight glaucous bloom, 18 to 21 inches long and broad at the middle, with the horny margin rolled in for 2 ins. or

3 ins. below the point, and finely toothed in the remaining portion. Flowers almost pure white, in a vast pyramidal panicle. This plant is hardy, but the fact is not generally known. It should be tried on well-drained slopes in good sandy loam. There are some varieties, of which *T. a. quadricolor* and *T. a. versicolor* have the leaves variously edged with green, yellow, and red. These fine variegated varieties are also very hardy, but as they are as yet far from common, it will be best to utilise them in the greenhouse or conservatory, or place them in the open air during summer. They look very pretty isolated on the grass, the pots plunged to the rim. S. America and W. Indies.

***Yucca angustifolia.**—A somewhat dwarf species, the whole plant, when in flower, not being more than 2 or 3 ft. high. The leaves are thick and rigid in texture, from 15 ins. to 18 ins. long and about ½ in. broad, of a pale sea-green colour, with numerous white filaments at the edges. The inflorescence is a simple raceme of white flowers slightly tinged with yellow. Till more plentiful this had better be grown in warm borders, in well-drained sandy loam. N. America.

***Yucca canaliculata.**—The leaves of this species are entire, *i. e.* neither toothed nor filamentous at the margin, and form a dense rosette on a stem which rises 1 or 2 ft. above the ground. Each leaf is from 20 ins. to 24 ins. long, and 2 ins. to 2½ ins. broad at the middle, very strong and rigid, and deeply concave on the face. The flowers are of a creamy white, in a large panicle 4 ft. to 5 ft. high. Fine for isolation or groups. Till more plentiful should be encouraged in favourable positions and on warm soils. Mexico.

***Yucca filamentosa.**—A very common and well-known species, with a much-branched panicle, 4 ft. to 6 ft. high, and apple-green leaves, from 15 ins. to 21 ins. long by 1½ ins. to 2 ins. broad at the middle, fringed at the edges with grey filaments 2 or 3 ins. long: the outer leaves spreading, the central ones erect or slightly recurved. This species varies very much when raised from seed: one variety (*concava*) has short, strong, broad leaves, with the face more concave than in the type; another variety (*maxima*) has leaves nearly 2 ft. long by 2½ ins. broad, with a panicle 7 ft. to 8 ft. in height. This species flowers with much vigour and beauty, and is well worth cultivating in every garden; not only in the flower-garden or pleasure-ground, but also on the rough rock-work, or any spot requiring a distinct type of hardy vegetation: and so is its fine though delicate variegated variety. All the varieties thrive best and flower most abundantly in peaty or fine sandy soil. N. America.

Yucca filamentosa.

***Yucca flaccida.**—A stemless species, somewhat resembling *Y. filamentosa*, but smaller, with a downy branching panicle 3 ft. to 4 ft. high. Foliage in close rosettes of leaves, 1½ ft. to 2 ft. long, by about 1½ in. broad at the middle, often fringed with filaments on the edges: the young ones nearly erect, the old ones abruptly reflexed at the middle, almost appearing as if broken. This gives such an irregular aspect to the tufts that it at once distinguishes this kind from any of the varieties of *Y. filamentosa*. It also flowers more regularly and abundantly than its relative, and is exceedingly well suited for groups of the finer hardy plants, for borders, or for being planted in large isolated tufts. N. America.

***Yucca glaucescens.**—A very free-flowering kind, with a panicle 3 ft. to 4 ft. high, the branches of which are short and very downy. Leaves sea-green, about eighteen inches long, with a few filaments on the margins. The flowers are of a greenish-yellow colour, and when in bud are tinged with pink, which tends to give the whole inflorescence a peculiarly pleasing tone. A very useful and ornamental sort—fine for groups, borders, isolation, or placing among low shrubs. N. America.

***Yucca gloriosa.**—A species of large and imposing proportions, with a distinct habit and somewhat rigid aspect. Flower-stem over 7 ft. high, much-branched, and bearing an immense pyramidal panicle, of large, almost pure white flowers. Leaves numerous, stiff, and pointed. One of the noblest plants in our gardens, and suitable for use in almost any position. It varies very much when grown from seed—a good recommendation, as the greater variety of fine form we have the better. The chief varieties in cultivation are *Y. g. longifolia*, *plicata*, *maculata*, *glaucescens*, and *minor*. The soil for this plant should be a rich deep loam. N. America.

***Yucca pendula.**—The best species perhaps, considering its graceful and noble habit, which is simply invaluable in every garden. It grows about 6½ ft. high, the leaves being at first erect and of a sea-green colour, afterwards becoming reflexed and changing to a deep green. Old and well-established plants of it standing alone on the grass are pictures of grace and symmetry, from the lower leaves which sweep the ground to the central ones that point up as straight as a needle. It is amusing to think of people putting tender plants in the open air, and running with sheets to protect them from the cold and rain of early summer and autumn, while perhaps not a good specimen of this fine thing is to be seen in the place. There is no plant more suited for planting between and associating with flower-beds. N. America.

***Yucca rupicola.**—A species somewhat resembling *Y. aloifolia*, with a stem from 4 ft. to 7 ft. high, and pale-green leaves 18 ins. to 20 ins. long, by 1 in. broad at the middle, almost erect and frequently twisted, the horny margin being broader and the teeth more distinct than in *Y. aloïfolia*. This is not much in cultivation as yet, and will probably be difficult to obtain for some time to come. N. America.

***Yucca Treculeana.**—This species is one of the most remarkable of the noble genus to which it belongs, from its habit, and especially from the dimensions to which its foliage attains. Like many plants of its family,

YUCCA PENDULA.

Hardy evergreen fine-foliaged Type.

young specimens differ considerably from those which have reached maturity. Thus, while the former have their leaves bent, generally inflected, the full-grown plants exhibit them erect, rigid, very long, and very straight. The stem of this plant is stout, about 10 ins. in diameter, furnished on all sides with leaves about 4 ft. long, straight, thick, deeply channeled, acuminate for a considerable length, and ending in a stiff, very sharp point, very finely toothed on the edges, which are of a brownish red and scarious. The flower-stalk is very stout, about 4 ft. long, much branched; the branches erect, from 1 ft. to 1 ft. 8 ins. long, bearing throughout their entire length flowers with long and narrow petals of a yellowish white, shining, and, as it were, glazed. It is a hardy and very vigorous plant. It is not rare to see on the Continent specimens of more than 6½ ft. in diameter. Fine for banks and knolls, placed singly, or for the boldest groups. It comes from Texas.

Zea Mays.—Were our climate a little warmer, we should find this noble grass one of the most ornamental, as well as useful, of our plants. But in countries where it is grown for food they would no more think of honouring it with a place in the garden than we should of planting the artichoke in our flower-beds, though far worse things are done every day. In this country, however, where maize is not to be seen as a field crop, a tuft of its tropical-looking blades has a good effect among the "subtropical plants." Of course it should only be tried in warm districts, and it should always have sunny and sheltered positions and rich soil. In light warm soils, deep, and with a free bottom, it generally thrives very well, if a foot or so of rich and rotten manure is placed beneath its roots. In some seasons it would here and there ripen seeds, and in all cases one could gather a few heads of "green corn." In warmer countries it is always best to sow maize in the open ground as soon as the frost permits; but in England it is better to raise it on a gentle hotbed in April, although occasionally it will succeed if sown out of doors. Gradually harden off the plants before they have made more than three or four little leaves, keeping them in a cool frame very near the glass, so as to keep them sturdy, and finally exposing them in the same position by taking the lights quite off. This course is perhaps the more desirable in the case of the variegated maize. In neither case should the plants be drawn up long in heat, as, if so, they will not thrive so well. The first few leaves the variegated kind makes are green, but they soon begin to manifest that striping which makes it as attractive as any variegated stove-plant we grow. *Cuzko* and *Caragua* are the largest and finest of the green varieties, and *gracillima* the smallest and most graceful of all the varieties of maize. They should be planted out about the middle of May.

The variegated or Japanese maize is a very remarkable and handsome variety, found by Mr. Hogg in Japan—that great country for variegated plants. Its beautiful variegation is reproduced true from seed, and it is almost an indispensable plant in the flower-garden, not growing so vigorously as the green kinds. It is particularly useful for intermingling with arrangements of ordinary bedding-plants, for vases, the outer margins of beds of subtropical plants, and like positions, where its variegation may be well seen, and where its graceful leaves will prove effective among subjects of dumpy habit. It should in all cases have light, rich, warm soil. It has a habit of breaking into shoots rather freely near the base of the central stem; and where it grows very freely, this should recommend it for planting in an isolated manner, or in groups of three or five, on the turf.

Yucca filamentosa variegata.

PART III.

SELECTIONS OF PLANTS FOR VARIOUS PURPOSES IN THE SUBTROPICAL GARDEN.

SUBTROPICAL GARDENING.

SELECTIONS OF PLANTS FOR VARIOUS PURPOSES.

A Selection of the very finest and most distinct Subtropical Plants, both hardy and tender, suited for use in the climate of Britain.

Acacia lophantha
Acanthus latifolius
Agave americana
„ „ variegata
Ailantus glandulosa
Aralia canescens
„ japonica
„ papyrifera
„ spinosa
Arundo conspicua
„ Donax
„ „ versicolor
Bambusa falcata
„ japonica
„ Simonii
„ viridi-glaucescens
„ edulis
Berberis Bealii
Beta cicla, var. chilensis
Caladium esculentum
Canna (in var.)
Chamærops excelsa
Crambe cordifolia
Corypha australis
Cycas revoluta
Dimorphanthus mandschuricus
Dicksonia antarctica
Dracæna indivisa
Echeveria metallica
Erythrina (in var.)
Ferdinanda eminens
Ferula (in var.)
Ficus elastica

Gynerium argenteum
Gunnera scabra
Gymnocladus canadensis
Helianthus orgyalis
Heracleum (in var.)
Melianthus major
Monstera deliciosa
Molopospermum cicutarium
Musa Ensete
Onopordon Acanthium
Paulownia imperialis
Phormium tenax
Poa fertilis
Polygonum cuspidatum
Polymnia grandis
Rheum (in var.)
Rhus glabra laciniata
Ricinus (in var.)
Seaforthia elegans
Solanum crinitipes
„ crinitum
„ macranthum
„ marginatum
„ robustum
„ Warscewiczii
Tupidanthus calyptratus
Uhdea bipinnatifida
Verbesina gigantea
Wigandia macrophylla
„ Vigieri
Yucca aloïfolia
„ canaliculata
„ gloriosa
„ pendula

A Selection of hardy perennials affording the finest effects in the Subtropical Garden.

Acanthus, in variety
Aralia edulis
„ nudicaulis
Astilbe rivularis
Arundo Donax
„ „ versicolor
Bambusa, in var.
Bocconia cordata
Carex paniculata
„ pendula
Carduus eriophorus
Carlina acaulis
Cassia marilandica
Centaurea babylonica
Crambe cordifolia
Datisca cannabina
Echinops ruthenicus
Eryngium alpinum
„ amethystinum
Gynerium argenteum
Gunnera scabra
Helianthus orgyalis

Hemerocallis fulva
Heracleum (in var.)
Inula Helenium
Melianthus major
Meum athamanticum
Molopospermum cicutarium
Morina longifolia
Panicum bulbosum
„ virgatum
Phytolacca decandra
Polygonum cuspidatum
Rhaponticum cynaroides
„ pulchrum
Rheum (in var.)
Statice latifolia
Tritoma (in var.)
Yucca (in var.)
Cynara Scolymus

A Selection of the finest tender Subtropical Plants that will succeed in our climate in summer.

Acacia lophantha
Agave americana
Aralia papyrifera
Asplenium Nidus-avis
Bambusa nigra
Bocconia frutescens
Brexia madagascariensis
Caladium esculentum
Canna (in var.)
Chamærops humilis
„ Palmetto
Cycas revoluta
Dahlia imperialis
Dracæna australis
„ cannæfolia
„ Draco
Echeveria metallica
Ferdinanda eminens
Ficus elastica
„ Chauvieri
Monstera deliciosa
Musa Ensete
Nicotiana virginica
„ wigandioides
Phormium tenax (hardy in the S. of England and Ireland)
Polymnia grandis
Ricinus (in var.)
Seaforthia elegans
Selinum decipiens
Solanum crinitipes
„ crinitum
„ macranthum
„ marginatum
„ robustum

Solanum Warscewiczii
Tupidanthus calyptratus
Uhdea bipinnatifida
Verbesina gigantea
Wigandia macrophylla
„ Vigieri
Zea Mays
„ „ variegata

A Selection of hardy Plants suited for isolation on the turf of the Flower-garden and Pleasure-ground.

Acanthus latifolius
„ longifolius
„ mollis
„ spinosissimus
„ spinosus
Aralia canescens
„ japonica
„ spinosa
Astilbe rivularis
Arundo conspicua
„ Donax
„ „ versicolor
Bambusa (in var.)
Bocconia cordata
Canna (hardier kinds)
Crambe cordifolia
Datisca cannabina
Dracæna indivisa (in the southern counties of England and Ireland)
Echinops ruthenicus
Elymus arenarius
„ condensatus
Eryngium alpinum
Eryngium amethystinum
Ferula (any kinds)
Gynerium argenteum
Gunnera scabra
Helianthus orgyalis
Hemerocallis flava
„ fulva and others
Heracleum eminens
Melianthus major
Molopospermum cicutarium
Morina longifolia
Osmunda regalis
Phormium tenax
Phytolacca decandra
Poa fertilis
Polygonum cuspidatum
Rheum Emodi (and other species and varieties)
Statice latifolia
Stipa pennata
Tritoma (any kind)
Yucca (any kind)

A Selection of Plants useful for the open air in summer and for embellishing the conservatory in winter.

Agave americana and vars. and other greenhouse species
Brexia madagascariensis
Chamærops excelsa
,, Fortunei
,, humilis
,, Palmetto
Cordyline indivisa
Cycas revoluta
Dracæna australis
,, cannæfolia
,, Draco
,, indivisa, and most of the other greenhouse kinds
Echeveria metallica
Ficus Chauvieri
,, elastica
Jubæa spectabilis
Monstera deliciosa
Musa Ensete
Phormium tenax, and vars.
Phœnix dactylifera and other greenhouse species
Seaforthia elegans
Tupidanthus calyptratus
Yucca aloifolia variegata, and vars.
Araucaria Bidwillii
,, Cookii
,, excelsa
,, Rulei
Areca sapida
Caryota urens
,, sobolifera
Corypha australis
Latania borbonica
Woodwardias
Half-hardy Palms, in var.

A Selection of hardy Plants of vigorous habit and distinct character suited for planting in semi-wild places in pleasure-grounds or near wood-walks.

Acanthus, in var.
Aralia canescens
,, edulis
,, nudicaulis
Aralia spinosa
Arum Dracunculus
Asclepias Cornuti
Asparagus Broussoneti

Astilbe rivularis
„ rubra
Arundo Donax
„ „ versicolor
„ Phragmites
Bambusa falcata
Bocconia cordata
Buphthalmum speciosum
Carex pendula
„ paniculata
Carduus eriophorus
Centaurea babylonica
Crambe cordifolia
„ juncea
Cucumis perennis
Datisca cannabina
Dipsacus sylvestris
Echinops ruthenicus
Elymus arenarius
Erianthus Ravennæ
Eryngium alpinum
„ amethystinum
Ferulas, in var.
Gunnera scabra
Helianthus orgyalis
„ Maximiliani
„ lætiflorus
„ occidentalis
„ rigidus
„ multiflorus
„ „ fl. pl.

Inula Helenium
Hemerocallis fulva
Heracleum, in var.
Lavatera arborea
„ thuringiaca
„ unguiculata
Hibiscus moscheutos
„ palustris
„ roseus
Althæa, in var.
Ligularia macrophylla
Molopospermum cicutarium
Morina longifolia
Mulgedium alpinum
„ Plumieri
Onopordon Acanthium
Pæonia, in var.
Panicum bulbosum
Papaver bracteatum
„ orientale
Petasites vulgaris
Phytolacca decandra
Poa aquatica
Polygonatum multiflorum
Polygonum cuspidatum
Rhaponticum cynaroides
„ pulchrum
„ scariosum
Rheum, in variety.
Rumex Hydrolapathum
Silphium, in var.

Silybum eburneum
,, marianum
Spiræa Aruncus
Statice latifolia
Tanacetum vulgare crispum
Thalictrum, in var.
Tritoma, in var.
Veratrum album
Verbascum, in var.
Yucca, in var.
Cynara Scolymus
Vernonia noveboracensis
Verbesina persicifolia
Rudbeckia digitata
,, laciniata
,, californica

A Selection of kinds that will best withstand wind.

Acacia Julibrissin
,, lophantha
Acanthus (all the kinds)
Agave americana
Ailantus glandulosa
Aralia canescens
,, japonica
,, spinosa
Artemisia annua
,, gracilis
Arundo conspicua
,, Donax
Astilbe rivularis
Bambusa falcata
Canna (in variety)
Carlina acaulis
Crambe cordifolia
Cycas revoluta
Datisca cannabina
Dracæna indivisa
Echinops ruthenicus
Elymus arenarius
Eryngium (in variety)
Ferula (in variety)
Ficus elastica
Gynerium argenteum
Kochia scoparia
Meum athamanticum
Molopospermum cicutarium
Osmunda regalis
Panicum bulbosum
,, virgatum
Phormium tenax
Phytolacca decandra
Poa fertilis
Polygonum cuspidatum
Rheum Emodi
Ricinus (in var.)
Tritoma (in var.)
Yucca (in var.)

Subtropical Plants to raise from seed.

Abutilon (in var.)
Acacia lophantha
„ Julibrissin
Acanthus, in var.
Amarantus, in var.
Aralia nudicaulis
„ papyrifera
„ japonica
„ spinosa
Artemisia annua
„ gracilis
Bocconia cordata
„ frutescens
Calla æthiopica
Canna, in var.
Cannabis sativa
Baptisia australis
„ exaltata
Beta cicla chilensis
Brassica oleracea crispa
Carduus eriophorus
Cassia marilandica
Centaurea, in var.
Cineraria acanthifolia
„ maritima
„ platanifolia
Chamæpeuce Cassabonæ
„ diacantha
Crambe cordifolia
Cyperus longus
Dahlia imperialis
Datura ceratocaula
Datisca cannabina
Dracæna, in var.
Echeveria metallica
Echinops ruthenicus
Dipsacus sylvestris
Erianthus Ravennæ
Erythrina, in var.
Eryngium alpinum
„ amethystinum
Ferdinanda eminens
Ferula, in var.
Astilbe rivularis
„ rubra
Galega officinalis
Gynerium argenteum
Gunnera scabra
Geranium anemonæfolium
Hedychium Gardnerianum
Helianthus orgyalis
Heracleum, in var.
Humea elegans
Inula Helenium
Kochia scoparia
Gourds
Latania borbonica
Lavatera arborea

Lobelia Tupa
Malva crispa
Melanoselinum decipiens
Melianthus major
„ minor
Meum athamanticum
Mulgedium alpinum
„ Plumieri
Musa Ensete
Nicotiana, in var.
Onopordon Acanthium
Panicum bulbosum
„ capillare
„ virgatum
Papaver bracteatum
Phormium tenax
Phytolacca decandra
Polymnia grandis
Rhaponticum cynaroides
„ pulchrum
Rheum, in var.
Ricinus, in var.
Salvia argentea
Seaforthia elegans
Silphium, in var.
Silybum eburneum
„ marianum
Solanum, in var.
Statice latifolia
Stipa pennata
Thalia dealbata
Thalictrum minus
Uhdea bipinnatifida
„ pyramidata
Verbascum Chaixii
Verbesina gigantea
Wigandia macrophylla
„ urens
„ Vigieri
Zea, in var.
Arundo conspicua

A Selection of annual and biennial Plants useful for the Subtropical garden.

[In this list annual plants grown for the beauty of the flower only are usually omitted.]

Adlumia cirrhosa
Amarantus, in var.
Argemone grandiflora
Artemisia annua
Artemisia gracilis
Atriplex hortensis ruber
Cannabis gigantea
„ sativa, and vars.

Chamæpeuce diacantha
„ Cassabonæ
Chenopodium Atriplicis
Cosmos, in var.
Gourds, in var.
Euphorbia variegata
Glaucium, in var.
Helianthus argyrophyllus
Kochia scoparia
Martynia lutea
Nicotiana Tabacum
„ virginica
Ricinus, in var.
Solanum erythrocarpum
Solanum Fontanesianum
„ racemigerum
Tagetes tenuifolia
Silybum eburneum
„ marianum
Chilian beet
Brassica oleracea crispa
Dipsacus sylvestris
Heracleum, in var.
Malva crispa
Onopordon Acanthium
„ tauricum
Zea, in var.

A selection of Flowers of various classes for association with Subtropical Plants.

[In this selection the dwarfer bedding-plants, etc., are omitted. Those selected are chiefly such as would bear more intimate association with fine-foliaged plants.]

Alstræmeria, in var.
Amaryllis Belladonna, and vars.
Gladioli, in great variety
Sparaxis pulcherrima
Lilium, in great variety
Agapanthus umbellatus, in the milder districts
Arum crinitum
Arum Dracunculus
Asclepias Cornuti
„ Douglasii
„ tuberosa
Calla æthiopica
Crinum capense
„ „ roseum
Erythrina, in var.
Funkia grandiflora

Pancratium illyricum
,, maritimum
Tropæolum speciosum
Acanthus longifolius
Achillea Eupatorium
,, Millefolium roseum
Aconitum, in var.
Ammobium alatum
Anchusa italica
Anemone japonica, and vars.
,, vitifolia
Antirrhinum, fine vars.
Asphodelus luteus
,, ramosus
Aster turbinellus
,, pyrenæus
,, discolor
,, ericoides
,, Novæ Angliæ
,, Novi Belgii
,, coccineus
,, Amellus
,, lævis, and any other tall and ornamental kinds
Campanula pyramidalis
,, persicifolia, and vars.
,, latifolia
,, macrantha

Coreopsis lanceolata
Crambe cordifolia
Delphinium, in great var.
Dictamnus Fraxinella
Digitalis purpurea, in var.
Echinops ruthenicus, and any other showy species
Epilobium angustifolium
,, ,, album
Stenactis speciosa
Erodium Manescavi
Eryngium alpinum
,, amethystinum, and other species
Eupatorium ageratoides
,, purpureum
Gaillardia, in var.
Galega officinalis
Hedysarum coronarium
Helenium atropurpureum
Helianthus multiflorus fl. pl.
Hemerocallis flava
,, fulva
,, disticha fl. pl., and others
Hesperis matronalis, fl. pl.
Iris pallida
,, De Bergii
,, ochroleuca
,, germanica, in var. and

any other large kind. Flowering early, they should be associated chiefly with hardy subjects
Lathyrus latifolius, and vars.
„ grandiflorus
„ tuberosus
Liatris, in var.
Lobelia Tupa, on well-drained, deep, and light soils. Tall herbaceous kinds in great variety
Lupinus polyphyllus, and vars. Largest annual kinds
Lychnis coronaria, in var.
Lythrum roseum superbum
„ virgatum
Michauxia campanuloides
Mirabilis Jalapa, in var.
Monarda, in var.
Morina longiflora
Œnothera, all the tall kinds
Pæonia, in great var.
Poppy, in var.
Pentstemon, in var.
Phlomis Herba-venti
„ tuberosa
„ Russelliana
Phlox, taller kinds, in great variety
Phygelius capensis, in warm districts
Polygonatum multiflorum, with Ferulas and other hardy things
Polygonum orientale
Potentilla, larger kinds in var.
Pyrethrum, choice double and single kinds in great var.
„ uliginosum
Rudbeckia Newmanni
„ hirta
Salvia patens
Saxifraga crassifolia
Schizostylis coccinea
Scabiosa caucasica
Scilla peruviana
Sedum spectabile
„ „ purpureum
Spiræa palmata
„ venusta
Statice latifolia
Stokesia cyanea, on warm soils in the south
Symphytum bohemicum
„ caucasicum
Thermopsis fabacea

Tradescantia virginica, and its varieties
Tigridia Pavonia, and other kinds
Tritoma, all the kinds
Trollius napellifolius
„ asiaticus, and others
Veronica, any tall herbaceous kinds, and in southern and mild districts the varieties of the evergreen New Zealand species
Vinca major, on the fringes of beds or groups of hardy kinds
Hollyhock, in var.
Dahlia, show, fancy, pompone, and bedding vars.
Verbascum Thapsus
„ Chaixii
Baptisia australis
Vernonia noveboracensis
Fuchsia, in var.
Datura ceratocaula
Abutilon, in var.
Ageratum, in var.
Petunia, in var.
Chrysanthemum, early-flowering kinds
Amarantus, in var.
Argemone grandiflora
Calliopsis, in var.
Cosmos bipinnatus purpureus
Echinacea angustifolia
„ atropurpurea
Hibiscus, any of the perennial American kinds
Malope, in var.
Matthiola, in var.
Scabiosa atropurpurea, in var.
French and African marigolds
Xeranthemum annuum, and vars.
Zinnia, in great variety
China aster, in var.
Brugmansia sanguinea

List of Plants for forming mixtures and carpets beneath Subtropical Plants.

Abronia umbellata
Acroclinium roseum
Ageratum mexicanum, and vars.

Alyssum maritimum
Anagallis indica
Athanasia, in var.
Brachycome iberidifolia
Calandrinia discolor
Calliopsis Drummondi
„ tinctoria
China aster, in var.
Centaurea
Centranthus macrosiphon
Clarkia, in var.
Clintonia, in var.
Collinsia, in var.
Convolvulus tricolor
Erysimum Peroffskianum
Eschscholtzia, in var.
Eucharidium concinnum
Eutoca viscida
Gaillardia picta
Gilia, in var.
Godetia, in var.
Gypsophila, annual kinds
Iberis coronaria
„ umbellata
Ionopsidium acaule
Lantana, in var.
Leptosiphon, in var.
Limnanthes Douglasii, in var.
Linum grandiflorum
Lobelia, the dwarf and annual kinds
Lupinus affinis
Mimulus, in var.
Myosotis palustris
„ dissitiflora
„ sylvatica
Nemesia versicolor
Nemophila, in var.
Nolàna, in var.
Omphalodes linifolia
Oxalis corniculata atropurpurea
Oxalis rosea
Petunia, in var.
Portulaca, in var.
Mignonette
Malcolmia maritima
Rhodanthe Manglesii, and vars.
Saponaria calabrica
Schizanthus, in var.
Silene pendula
Sphenogyne speciosa
Tropæolum, the bedding vars.
Verbena, in var.
Viola cornuta
„ lutea
„ odorata
Viscaria oculata
Whitlavia grandiflora
Tradescantia zebrina

Saxifraga, the mossy section
Gnaphalium lanatum
Panicum variegatum
Lycopodium denticulatum

Trees and Shrubs of remarkable foliage suited for the Subtropical garden.

[The trees of this Selection will for the most part display much greater beauty and size of foliage if kept in a dwarf simple-stemmed condition by being cut down every year. Conifers are, of course, excepted.]

Hydrangea quercifolia
Comptonia asplenifolia
„ Lindleyana
Paulownia imperialis
Acacia dealbata
„ Julibrissin
„ lophantha
(These are only suited for warm parts of the southernmost counties)
Ailantus glandulosa
Aralia canescens
„ japonica
„ spinosa
Bambusa falcata and others
Berberis Bealii and others
Chamærops excelsa
Dracæna indivisa
Gymnocladus canadensis
Lavatera arborea
Melianthus major
Rhus glabra laciniata and others
Yucca (in var.)
Conifers (in var., small specimens of the most graceful kinds)
Ampelopsis (climbers)
Aristolochia Sipho (climber)
Ficus Carica
Fraxinus excelsior crispa
Magnolia macrophylla
Salisburia adiantifolia
Hedera Regnieriana
Carya alba
„ olivæformis
Catalpa syringæfolia
Pterocarya caucasica
Robinia hispida
Rubus biflorus
„ laciniatus
Colletia (in var.)
Gleditschia (young plants)

Kœlreuteria paniculata
Robinia Pseud-acacia umbraculifera
Tamarix, in var.
Vines (American species)
Juglans cinerea
„ regia
„ „ laciniata

A Selection of Conifers for association with flower-garden Plants.

Abies pygmæa
Araucaria imbricata
Arthrotaxus cupressoides (selaginoides)
„ laxifolia
Biotia cupressiformis
„ nana
„ orientalis elegantissima
„ orientalis variegata aurea
Cephalotaxus drupacea
Chamæcyparis sphæroidea variegata
„ sphæroidea viridis
Cryptomeria elegans
Cupressus Lawsoniana
„ „ nana
„ „ variegata
„ „ erecta viridis
Cupressus nutkaensis
Dacrydium glaucum
Juniperus chinensis
„ „ variegata
„ excelsa stricta
„ fragrans
„ hibernica
„ tamariscifolia
„ „ variegata
„ virginiana viridis pendula
Retinospora ericoides
„ leptoclada
„ lycopodiodes
„ obtusa
„ „ aurea
„ „ compacta
„ pisifera
„ „ alba variegata
„ „ aurea

Taxus baccata elegantissima	Thuja gigantea
" " variegata	" nana
Thuja aurea	Thujopsis dolabrata
	" lætevirens

Conifers most suited for the surroundings of the flower-garden and pleasure-ground—kinds which, though noble and graceful as can be in many instances, are yet too large for anything but the framing of the picture, so to speak.

Abies Douglasii	Picea Parsonsi
" Engelmanni	" Pinsapo
" Menziesii inverta	Pinus Cembra
" Hookeriana	" monticola
" orientalis	" insignis (where it thrives)
Cephalotaxus Fortunei	Sciadopitys verticillata
Juniperus virginiana glauca	Thuja gigantea (true)
" " thurifera	" plicata
Libocedrus tetragona	" pyramidalis
Picea amabilis (magnifica)	Thujopsis Standishii
" nobilis	Sequoia gigantea
" Nordmanniana	" sempervirens

A Selection of Gourds.

Amongst the most beautiful are the Turk's Cap varieties, such as Grand Mogul, Pasha of Egypt, Viceroy, Empress, Bishop's Hat, etc.; the Serpent Gourd, Gooseberry Gourd, Hercules' Club, Gorilla, St. Aignan, Mons. Fould, Siphon, Half-moon, Giant's Punchbowl, and the Mammoth, weighing from 170lb. to upwards of 200lb.; while amongst the miniature varieties the Fig, Cricket-ball, Thumb, Cherry, Striped Custard, Hen's-egg, Pear, Bottle, Orange, Plover's-egg, etc., are very pretty examples, and very serviceable for filling vases, etc. All these are well adapted to the climate of England, and there are many others equally suitable—a fact sufficiently indicated in one collection shown by Mr. W. Young, which consisted of 500 varieties, all English grown, the greater number of which were sown where grown, and came to maturity without the assistance of glass or any other protection. The ground being manured and dug one spit deep, the seed was sown the second week in May, and from first to last many of the plants had no water supplied to them through the season. Others, by way of experiment, had it in various quantities—the more water was given, the larger, the freer, and the better the produce. Sowing in a frame at the end of April, and exposing them to the free air during the day so as to prevent them being drawn, and then removing the frame altogether to harden them off before planting out, would be the best way to secure an early growth of gourds. Sowing in the open ground under hand-lights would also do, but not so well.

Ornamental Grasses.

Agrostis nebulosa
Arundo conspicua
„ Donax
„ „ versicolor
„ festucoides
„ Phragmites
Bambusa, in var.
Elymus arenarius
„ condensatus
Erianthus Ravennæ
Gynerium argenteum, and its vars.

Calamagrostis argentea
Poa aquatica
„ fertilis
Saccharum ægyptiacum
„ cylindricum
„ Maddenii
Stipa pennata
Zea Mays
Andropogon argenteus
„ bombycinus
„ formosus
„ Sorghum
„ strictus
„ squarrosus
Chloropsis Blanchardiana
Gymnothrix latifolia
Holcus saccharatus
Erianthus strictus
„ violascens
Chloris myriostachys
Panicum bulbosum
„ altissimum
„ capillare
„ miliaceum
„ virgatum
„ maximum
„ palmifolium
„ gongyloides
Panicum violaceum
Penicillaria spicata
Sorghum cernuum
„ halepense
„ melanocarpum
„ nankinense
„ tataricum
Tripsacum monostachyum
„ dactyloides
Milium nigricans
„ multiflorum
„ effusum
Bromus brizopyroides
Briza gracilis
„ geniculata
„ maxima
„ rufiberbis
Hordeum jubatum
Pennisetum longistylum
Piptatherum multiflorum
Agrostis spica-venti
Setaria germanica
Stipa capillata
Chascolytrum erectum
Leptochloa gracilis
Agrostis Steveni
Echinochloa Zenkowski
Paspalum elegans

List of Ferns that may be grown with advantage away from the fernery proper.

[Even should any of these thrive better in shade, it is usually easy to secure this for them in groups by wood-walks.]

Adiantum pedatum
Asplenium Filix-fœmina and vars.
Dennstœdtia punctilobula
Diplazium thelypteroides
Lastrea Filix-mas and vars.
,, Goldieana
,, ,, assurgens intermedia
,, marginalis
,, noveboracensis
,, atrata
,, erythrosora
,, opaca
,, Standishii
Lomaria magellanica, in warm shady places
Onoclea sensibilis
Osmunda cinnamomea
,, Claytoniana
,, gracilis
,, regalis
Osmunda regalis cristata
,, spectabilis
Polypodium hexagonopterum
Polypodium Phegopteris
Polystichum acrostichoides
,, aculeatum
,, angulare
,, vestitum venustum
Pteris aquilina
Scolopendrium vulgare and vars.
Struthiopteris germanica
,, pennsylvanica
Woodwardia areolata
,, aspera
,, japonica
,, orientalis
,, radicans
Cyrtomium caryotideum
,, falcatum

List of hardy aquatics and bog-plants of bold and distinct habit suitable for grouping n the margins of lakes, etc.

Nuphar lutea	Poa aquatica
„ advena	Arundo Phragmites
„ pumila	Cyperus longus
Nymphæa alba	Cladium Mariscus
„ odorata	Pontederia cordata
Menyanthes trifoliata	Acorus Calamus
Equisetum Telmateia	Iris Pseudacorus
Rumex Hydrolapathum	Alisma Plantago
Typha angustifolia	Orontium aquaticum
„ latifolia	Lysimachia thyrsiflora
Carex pendula	Lythrum Salicaria
„ paniculata	Epilobium hirsutum
„ Pseudocyperus	Calla æthiopica
Scirpus lacustris	„ palustris
Butomus umbellatus	Hippuris vulgaris

THE END.

LONDON: PRINTED BY WILLIAM CLOWES AND SONS, STAMFORD STREET AND CHARING CROSS.

R

Printed in the United States
By Bookmasters